100道
鹹甜點心
Dim Sum
& Snacks

序

懂得寵愛自己，讓自己充滿正面陽光的積極力量，才有能力寵愛身邊的人。

讓大家幸福、歡笑、健康，是我的「使命」。

為了帶來更棒的影響力，我費盡心思，嚴格挑選每項食材，按完美比例調配下，誠意撰寫了鹹味美點及輕食甜蜜的食譜，願你能輕鬆愉快地享受製作家常美食的樂趣。

西諺有云：「停下腳步，聞聞花香」（stop and smell the roses），偶爾停下腳步與家人摯友相伴，以茶配美食及甜點，徜徉於美好的時光，品嘗人生況味吧！

感謝天父，承蒙圓方出版社的邀請及簡小姐的協助，看到出版的第三本食譜，給我無比的自信心，更懂得愛自己、愛別人！

雖然生活忙得喘不過氣來，卻能完成此書，那動力源自愛護我的家人～大兒子柏年、媳婦 Polly 及小兒子阿冠；廚藝班及陪月班的學生；教會的弟兄姊妹；李國夫婦、黃耀榮先生、蘇憲先生、胡仕忠先生及戚珍雁小姐等。他們的關心鼓勵與陪伴，使我跟大家分享了美好的事物。

最後，謹以此書向天父獻上感恩，也以此書紀念在天父懷抱裏的先父葉志超先生。

胡影儀

目錄
Contents

鹹．點心

少變化，多口味

鍋貼 68
Shanghainese pot stickers

粢飯 70
Shanghainese sticky rice roll

鮮蝦餃 **Video** 72
Har Gow (Cantonese shrimp dumpling)

少變化，多口味

雲耳鮮蝦餃 76
Steamed cloud ear shrimp dumpling

菠菜皮鮮蝦餃 78
Har Gow with spinach skin

上湯菜肉餃 80
Bok choy and pork dumplings in stock

鹹·小吃

香滑蝦米腸粉 **Video** 82
Rice noodle roll with dried shrimps

碗仔翅 85
Imitation shark's fin soup

越南春卷 88
Vietnamese spring rolls

鹹湯丸 91
Dace balls and glutinous rice balls in stock

自製蜜汁豬肉乾 94
Home-made honey-glazed pork jerky

白雲豬手 96
Cold pork trotters in pickle brine

葱油餅 98
Spring onion pancake

少變化，多口味

香荔酥卷 101
Beancurd skin roll with taro filling

荔茸葱油餅 102
Taro spring onion pancakes

蝦肉木耳腐皮夾 104
Beancurd skin sandwich with
minced shrimp and wood ear

醃雲耳蘿蔔皮 106
Radish skin and cloud ear fungus pickles

潮州煎菜頭粿 109
Chaozhou pan-fried dried radish cakes

香醬汁雞翼 112
Deep-fried chicken wings in
nutty fruity soy sauce

魷魚鬚雜菜天婦羅 114
Tempura squid tentacles and assorted veggies

臘味蘿蔔糕 116
Radish cake with preserved sausage and pork

五香蛋散仔 119
Deep-fried mini twisted pretzels

芝麻蝦 122
Sesame shrimp toast

雞絲粉皮 124
Ribbon glass noodles with shredded chicken

咖喱角 126
Samosa

煙肉韭菜薄餅 128
Bacon and chive pancakes

豉油浸沖菜 130
Soy-marinated kohlrabi

錦滷雲吞 132
Deep-fried wontons in sweet and sour sauce

蒸夾心雞蛋糕 196
Steamed cake with lotus seed paste filling

蜂蜜綠茶凍糕 198
Honey green tea jelly

薑汁撞奶 200
Ginger milk custard

椰汁紅豆糕 202
Coconut milk cake with red bean filling

蒸香蕉蛋糕 205
Steamed banana cake

杞子桂花糕 208
Osmanthus jelly with Qi Zi

雙色椰汁糕 210
Coconut cream mousse

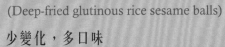

甜·小吃

鮮芒果椰汁河粉 212
Coconut milk noodles in mango puree

豆沙煎軟糍 215
Sticky rice pancakes with red bean filling

糖不甩 218
Glutinous rice balls with peanut sesame topping

笑口棗 220
Deep-fried crunchy sesame balls

炸番薯煎堆 222
Sweet potato Jian Dui
 (Deep-fried glutinous rice sesame balls)

少變化，多口味

擂沙煎堆 226
Jian Dui with black sesame filling

南瓜煎堆 228
Pumpkin Jian Dui

甜薄罉 230
Chinese crepe with sweet filling

豆沙角 232
Deep-fried red bean dumplings

朱古力雞蛋仔 234
Chocolate egg waffles

椰汁糯米糕 238
Coconut glutinous rice cake with red bean filling

紅豆缽仔糕 240
Put Chai Ko with red beans
 (Steamed rice cake in bowls)

酥炸角仔 242
Yau Gok (Cantonese deep-fried oil dumplings)

豆沙鍋餅 246
Deep-fried crispy pancake with red bean filling

擂沙湯丸 248
Glutinous rice balls in ground peanut flour

蓮茸西米角 250
Sago dumplings with lotus seed paste filling

蒸南瓜餅 254
Steamed pumpkin cake

拔絲香蕉 256
Candied banana fritters

高力豆沙 259
Soufflé egg white balls with red bean filling

Basic utensils for dim sum

量杯 Measuring cup

秤量液體及粉類材料，必須以眼睛平視刻度才準確。本書計量粉料時，以量杯作為工具，方便一般沒有磅秤的家庭；液體以量杯為計量用具。

It is for the measurement of liquid and dry ingredients (all flour and starches). You must read it at eye level. All dry ingredients in this book are measured by measuring cups if you have no scale at home.

量匙 Measuring spoons

一般分為1湯匙、1茶匙、1/2茶匙、1/4茶匙及1/8茶匙，舀取粉料後刮平為準。

They come in 1 tbsp, 1 tsp, 1/2 tsp, 1/4 tsp and 1/8 tsp. When you use measuring spoons to measure dry ingredients, always level off any excess with a dough scraper.

刮板 Dough scraper

可拌麵糊、搓揉或切割麵糰，甚至將黏着的材料刮淨。

It is used to stir batter, and to knead or cut dough. You can also use it to scrape off any ingredient that sticks to a surface.

白洋布 Muslin cloth

即白色的棉布,可用作製腸粉或蒸糯米等,一般布行有售,價錢便宜。

It is white cotton cloth used for making rice rolls or steaming glutinous rice. You can get it from fabric stores. It costs very little.

擀麵棒 Rolling pin

壓製麵糰的工具,可壓成不同厚薄的麵皮,宜預備粗或幼擀麵棒各一支。

It is used to roll out the dough into different thicknesses. It's best to stock one thick rolling pin and a thin one.

木製印花壓模 Wooden decorated mould

製作軟糕時使用,可壓出細致的花紋。

It is used to make soft cakes. You can press the dough into the mould to give the cake intricate patterns.

蒸籠 Steamer

用於蒸製點心、包子,以竹蒸籠較佳,有大小不同尺碼。一般的不銹鋼蒸籠無法吸收蒸煮時的水蒸氣,影響製成品賣相。

This is where buns and dim sum are steamed. Bamboo steamers usually work better than stainless steel ones and they come in different sizes. Stainless steel steamers don't absorb the steam in the steaming process and that might adversely affect the presentation.

揉麵糰的技巧

- 搓餃子皮不似製包子方法，毋須發酵步驟，但卻留意搓麵糰時水溫對質感的影響。水溫 20℃ 令麵糰產生彈性，有咬勁，可製成水餃皮；沸水 70℃ 以上令麵糰失去彈性，只宜做成鍋貼或蒸餃等。

- 先用手的熱力搓溶砂糖，才混和麵粉，可搓成均勻的麵糰。

- 搓麵糰時，撒些麵粉於桌面以免黏手，此稱為「手粉」。

- 麵糰要充份揉好，搓至表面光滑，別出現水量太多黏手的情況。

- 環境的溫度對發酵麵糰很重要，最適宜的溫度是 27 至 30℃。

- 麵糰搓好後，宜用濕布蓋着，以免麵糰太乾，失去柔軟度。

- 擀麵皮時，用右手不停滾動擀麵棒，左手轉動麵皮，令擀出來的麵皮均勻。包裹餡料的中間位置宜厚一點，皮邊可薄一點，以免麵皮穿破。

Dough kneading techniques

- As opposed to making buns, you don't need to ferment the dough when you make dumplings. Yet, you still have to pay attention to how water temperature affects the texture of the dumpling skin. Adding water at 20°C gives some kind of springiness and chewiness to the skin which is desirable in soup dumplings. On the other hand, adding hot water at 70°C or above would make the skin less resilient. Such dough is mostly used in pot-stickers or steamed dumplings.

- Use your hands to rub and melt the sugar first. Then stir in the flour. This is the key to consistently mixed dough.

- When you knead the dough, flour the counter and your hands.

- Make sure the dough is sufficiently kneaded and it should be smooth on the surface. It's too wet if it sticks to your hand.

- Ambient temperature is important to fermentation of the dough. The best rising temperature is between 27°C and 30°C.

- After kneading the dough, cover it in damp cloth. It might lose its softness and resilience if it dries up.

- When you roll out the dough, roll the rolling pin with your right hand while turning the dough with your left hand. That's how you make the dough equally thick all over. The centre of the rolled out dough can be slightly thicker while the edge should be thinner. The dumpling skin is less likely to break that way.

自家製。無添加豬油

Home-made lard from scratch

雖說豬油不健康，但製作某些點心時，我也會用少許豬油，保留點心精髓之餘，也令味道更濃郁。可以的話，跟我自製無添加豬油，為美味與健康找個平衡點。

材料 Ingredients

肥豬肉半斤
300 g fatty pork

做法 Method

1. 肥豬肉洗淨，切件。
2. 燒熱半碗水，放入肥豬肉及鹽半茶匙，加蓋，調小火煮至水分收乾。
3. 揭開蓋，以極小火炸至肥豬肉出油及乾身，盛起豬油，冷藏可貯存數月。

1. Rinse the pork and cut into pieces.

2. Boil 1/2 bowl of water. Put in the pork and 1/2 tsp of salt. Cover the lid. Turn to low heat and simmer until water dries up.

3. Open the lid and fry the fatty pork over the lowest heat until it gives oil and turns crispy. Set aside the lard. It stores well in the fridge for a few months.

✎ 小煮意 ✎

- 半斤肥豬肉約得一飯碗豬油。
- 用極小火炸出來的豬油呈白色，令做出來的點心色澤白淨，賣相佳。
- 炸油後的豬油渣可拌飯麵、炒菜等，香氣撲鼻。

- 300 g of fatty pork should yield about 1 bowl of lard.

- You must fry the pork over the lowest heat so that the lard remains clear without being browned. The dim sum made with it will also remain white in colour and look more pleasing.

- The pork turns into cracklings after being fried and all grease is removed. You can stir cracklings in with noodles or use them to in stir-fried vegetables. They add a meaty flavour to your food.

Using ready-made dumpling skin to save time

愛入廚的你，總希望製成品百分百出自雙手的努力，那份美味，包含着成功感。餃皮，可自行調勻水及麵粉搓成，當中滿有樂趣。若無暇自製餃皮，可購買製麵店的雲吞皮或餃子皮，變化出多款點心美食。

- 正方形餃子皮：皮略厚，大塊，包成餃子最恰當。

- 燒賣皮：用全蛋及高筋麵粉搓成，呈黃色，質感薄且帶韌性，可製燒賣或蝦角，但鹼水味重。

- 圓形餃子皮：薄身，容易穿破，宜包乾性餡料。

- 黃色雲吞皮：皮薄，不宜包裹含水份多的蔬菜餡料，多包製肉類及蝦等；炸製後也可做成錦滷雲吞。

- Square dumpling skin: It's thicker than the rest and comes in big pieces. It's best for making soup dumplings.

- Shaomai skin: It's the yellow dumpling skin made from whole eggs and bread flour. It's thin but quite resilient. You can use it for Shaomai or deep-fried shrimp dumplings. Yet, it has a rather strong taste of lye.

- Round dumpling skin: It's thin and tends to break easily. It should only be used for dry filling ingredients.

- Yellow wanton skin: It is paper thin and you should not use it to wrap ingredients with high water content, such as vegetables. You can use it for wrapping meat or shrimps, and deep-fried wantons in sweet and sour sauce.

鮮蝦燒賣皇 Shrimp Shaomai

材料 Ingredients

瘦肉 4 兩
肥豬肉 1 兩
蝦肉 4 兩
冬菇 6 朵
中蝦 10 隻（去殼）
蟹籽適量
豬油 1 湯匙
圓形燒賣皮 4 兩
150 g lean pork
38 g fatty pork
150 g shelled shrimps
6 dried shiitake mushrooms
10 medium shrimps (shelled)
crab roe
1 tbsp lard
150 g round Shaomai skin

調味料 Seasoning

豬油 1 湯匙
鹽半茶匙
糖 3/4 茶匙
胡椒粉 1/4 茶匙
麻油 1/4 茶匙
生抽 1 茶匙
紹酒 1 茶匙
浸冬菇水 1 湯匙
生粉 1 湯匙
蛋液 1 湯匙
1 tbsp lard
1/2 tsp salt
3/4 tsp sugar
1/4 tsp ground white pepper
1/4 tsp sesame oil
1 tsp light soy sauce
1 tsp Shaoxing wine
1 tbsp water in which shiitake mushrooms
are soaked
1 tbsp caltrop starch
1 tbsp whisked egg

Video

❧ 小煮意 ❧

- 餡料拌入肥豬肉，蒸出來的燒賣更香更滑。

- 肥豬肉切粒拌勻即可，不可剁爛，否則釋出大量油分。

- 瘦肉則以人手剁爛的口感較佳。

- 圓形燒賣皮於售賣點心的批發店有售，若然難以購買，可將黃色雲吞皮裁剪成圓形包裹。

- 燒賣包裹後噴上水，餡料不容易散開，而且更濕潤。

- The diced fatty pork adds both velvety texture and a meaty aroma to the Shaomai.

- The fatty pork should only be diced not chopped before stirred into the filling. Otherwise, it may give too much oil and make the Shaomai too greasy.

- On the other hand, the lean pork should be chopped with a knife for better mouthfeel. I don't recommended getting ground pork or grinding it in a machine.

- Round Shaomai skin is available from wholesalers of Dim Sum supplies. If you can't get it, just use yellow Wanton skin and cut it into a round disc.

- Spraying water on Shaomai before steaming helps them hold their shape better. The Shaomai won't dry out in the steaming process either.

做法 Method

1. 冬菇浸軟,去蒂,切絲,冬菇水留用。

2. 肥豬肉煮熟,沖淨,切幼粒;瘦肉剁爛或切粒;蝦肉用生粉及鹽各半茶匙醃 5 分鐘,洗淨,用乾布壓乾水分,切粒;中蝦用鹽、胡椒粉及麻油各少許略醃。

3. 將所有材料拌勻(中蝦除外),下調味料順一方向攪拌,撻至起膠,放入雪櫃 待 15 分鐘。

4. 將燒賣皮放於大拇指與食指之間的圓環上,包入餡料略壓,用手指輕捏,鋪上 中蝦。

5. 燒賣排入蒸籠內,噴水,以大火蒸 10 分鐘,以蟹籽裝飾即可。

1. Soak shiitake mushrooms in water until soft. Cut off the stems. Shred them. Set aside the water you use to soak the mushrooms for later use.

2. Boil the fatty pork in water until done. Rinse and finely dice it. Set aside. Finely chop the lean pork or finely dice it. Mix the shelled shrimps with 1/2 tsp of caltrop starch and 1/2 tsp of salt. Leave them for 5 minutes. Rinse and squeeze dry with a dry towel. Dice them. Mix the medium shrimps with salt, ground white pepper and sesame oil.

3. Put all ingredients into a mixing bowl (except medium shrimps). Add seasoning and stir in one direction until sticky. Leave the filling in the fridge for 15 minutes.

4. Make a circle with your thumb and first finger. Put a piece of Shaomai skin into the round space. Put some filling on the Shaomai skin while pressing gently so that the Shaomai sinks into the space. Put a medium shrimp on top. Repeat this step until all ingredients are used up.

5. Put the Shaomai into a steamer. Spray water on them. Steam over high heat for 10 minutes. Put on crab roe. Serve.

糯米燒賣

少變化，多口味

小煮意

- 用蒸籠蒸煮糯米，比用電飯煲更有口感。
- 糯米餡料必須充份待涼，才包入燒賣皮內。

- Steamed the glutinous rice retains a slightly chewy texture. It tastes better than that cooked in a rice cooker.
- The glutinous rice filling must be completed cooled before stuffed into the Shaomai skin.

Glutinous rice Shaomai

材料 Ingredients

糯米 1 1/2 杯
冬菇 4 朵（浸軟，切粒）
蝦米 4 湯匙（切粒）
臘腸 1 條（切粒）
臘肉半條（切粒）
葱白 3 條（切粒）
圓形燒賣皮 3 兩

1 1/2 cup glutinous rice
4 dried shiitake mushrooms (soaked in water until soft; diced)
4 tbsp dried shrimps (diced)
1 Chinese dried pork sausage (diced)
1/2 piece Chinese dried pork belly (diced)
3 sprigs spring onion (white part only, diced)
113 g round Shaomai skin

調味料 Seasoning
生抽 1 湯匙
老抽 1 茶匙
糖 1 茶匙
紹酒、麻油及胡椒粉各少許
上湯 2 湯匙
1 tbsp light soy sauce
1 tsp dark soy sauce
1 tsp sugar
Shaoxing wine
sesame oil
ground white pepper
2 tbsp stock

做法 Method

1. 糯米用水浸 4 小時，洗淨，瀝乾，下鹽半茶匙及油少許，蒸約半小時，待涼。
2. 燒熱鑊，下油 3 湯匙，加入冬菇粒、蝦米、臘腸粒、臘肉粒及葱白炒勻，下調味料拌勻，放入糯米飯炒勻，上碟，待涼備用。
3. 將燒賣皮放於大拇指與食指間的圓環上，包入糯米餡略壓，用手指輕捏。
4. 燒賣排入蒸籠內，餡面掃上少許油，噴水，以大火蒸 5 分鐘即成。

1. Soak the glutinous rice in water for 4 hours. Rinse well and drain. Add 1/2 tsp of salt and a dash of oil. Steam for 30 minutes. Let cool.

2. Heat a wok and add 3 tbsp of oil. Stir fry shiitake mushrooms, dried shrimps, dried pork sausage, dried pork belly and white part of spring onion. Add seasoning and mix well. Put in the steamed glutinous rice. Mix well and set aside to let cool. This is the filling.

3. Make a circle with your thumb and first finger. Put a piece of Shaomai skin into the round space. Put some glutinous rice filling on the Shaomai skin while pressing gently so that the Shaomai sinks into the space. Repeat this step until all ingredients are used up.

4. Put the Shaomai into a steamer. Brush oil on the glutinous rice filling. Spray water on them. Steam over high heat for 5 minutes. Serve.

Pork Shaomai

材料 Ingredients
瘦肉 4 兩
肥豬肉 1 兩
蝦肉 6 兩
冬菇 4 朵
青豆 2 兩
豬油 1 茶匙
圓形燒賣皮 4 兩
150 g lean pork
38 g fatty pork
225 g shelled shrimps
4 dried shiitake
mushrooms
75 g green peas
1 tsp lard
150 g round Shaomai
skin

調味料 Seasoning

豬油 1 茶匙	1 tsp lard
鹽半茶匙	1/2 tsp salt
糖半茶匙	1/2 tsp sugar
胡椒粉 1/4 茶匙	1/4 tsp ground white pepper
麻油 1/4 茶匙	1/4 tsp sesame oil
生抽 1 茶匙	1 tsp light soy sauce
紹酒半茶匙	1/2 tsp Shaoxing wine
浸冬菇水 1 湯匙	1 tbsp water in which shiitake mushrooms are soaked
生粉 1 湯匙	1 tbsp caltrop starch
蛋液 1 湯匙	1 tbsp whisked egg

做法 Method

1. 冬菇浸軟，去蒂，切絲，冬菇水留用。
2. 肥豬肉煮熟，沖淨，切幼粒；瘦肉剁爛或切粒；蝦肉洗淨，用乾布壓乾水分，切粒。
3. 將所有材料拌勻（青豆除外），下調味料順一方向攪拌，撻至起膠，放入雪櫃待 15 分鐘。
4. 將燒賣皮放於大拇指與食指之間的圓環上，包入餡料略壓，用手指輕捏，表面以青豆裝飾。
5. 燒賣排入蒸籠內，噴水，以大火蒸 10 分鐘即可。

1. Soak shiitake mushrooms in water until soft. Cut off the stems. Shred them. Set aside the water you use to soak the mushrooms for later use.

2. Boil the fatty pork in water until done. Rinse and finely dice it. Set aside. Finely chop the lean pork or finely dice it. Rinse the shelled shrimps. Wipe and squeeze dry with a dry towel. Dice them.

3. Put all ingredients into a mixing bowl (except green peas). Add seasoning and stir in one direction until sticky. Leave the filling in the fridge for 15 minutes.

4. Make a circle with your thumb and first finger. Put a piece of Shaomai skin into the round space. Put some filling on the Shaomai skin while pressing gently so that the Shaomai sinks into the space. Put a green pea on top. Repeat this step until all ingredients are used up.

5. Put the Shaomai into a steamer. Spray water on them. Steam over high heat for 10 minutes. Serve.

Dumpling soup
with shrimp filling 鮮蝦水餃

皮料 Ingredients

白色水餃皮半斤
300 g white dumpling skin

餡料 Filling

梅頭豬肉 5 兩
蝦肉 6 兩
筍肉 2 兩
木耳絲 4 湯匙
冬菇 6 朵
韭黃 3 兩
雞蛋 1 個
190 g pork shoulder butt
225 g shelled shrimps
75 g bamboo shoot (leaves
removed and trimmed, heart only)
4 tbsp shredded wood ear fungus
6 dried shiitake mushrooms
113 g yellow chives
1 egg

蒸冬菇調味料
Seasoning for
shiitake mushrooms

生抽、糖、油及紹酒各 1 茶匙
浸冬菇水 2 湯匙
1 tsp light soy sauce
1 tsp sugar
1 tsp oil
1 tsp Shaoxing wine
2 tbsp water in which
shiitake mushrooms are soaked

醃料 Marinade

鹽 1 茶匙
生粉 1 湯匙
糖半茶匙
生抽、麻油、胡椒粉各少許
1 tsp salt
1 tbsp caltrop starch
1/2 tsp sugar
light soy sauce
sesame oil
ground white pepper

上湯料 Stock

上湯 5 杯
鹽 1 1/2 茶匙
紹酒及麻油各少許
5 cups stock
1 1/2 tsp salt
Shaoxing wine
sesame oil

做法 Method

1. 冬菇浸軟，去蒂，加入冬菇調味料蒸 15 分鐘，切絲。

2. 豬肉切粒、剁碎；蝦肉去腸，用生粉及鹽各半茶匙醃 5 分鐘，洗淨及抹乾，切粒。

3. 筍肉飛水，瀝乾水分，切絲；木耳飛水，切絲；韭黃切粒。

4. 將所有餡料拌勻（韭黃及雞蛋除外），加入醃料及蛋黃攪拌。

5. 水餃皮放入餡料，包捏成餃形，以蛋白塗抹皮邊，壓實，放入滾水內煮至浮起，再下清水半碗煮滾至熟透，置於碗內。

6. 燒滾上湯，傾入碗內，灑上韭黃即可享用。

1. Soak the shiitake mushrooms in water until soft. Cut off the stems. Add seasoning for shiitake mushroom and mix well. Steam for 15 minutes. Finely shred them.

2. Dice the pork and then finely chop it. Devein the shelled shrimps. Add 1/2 tsp of caltrop starch and 1/2 tsp of salt. Leave for 5 minutes. Rinse and wipe dry. Dice them.

3. Blanch the bamboo shoot in boiling water. Drain and shred it. Blanch the wood ear fungus in boiling water. Finely shred it. Dice the yellow chives.

4. Put all ingredients (except yellow chives and eggs) into a mixing bowl. Stir to mix well. Then add marinade and egg yolk. Stir again. This is the filling.

5. Put some filling on a piece of dumpling skin. Brush some egg white on the edge. Fold into a dumpling shape. Press the seam firmly to secure. Boil the dumplings in water until they float. Pour 1/2 bowl of water and bring to the boil. Transfer into a bowl.

6. Boil the stock. Pour into the bowl over the dumplings. Sprinkle with yellow chives on top. Serve.

小煮意

想吃出爽口彈牙的蝦肉，蝦去殼去腸後，加入生粉及鹽各 1 茶匙醃 5 分鐘，洗淨及抹乾後即可烹調。

To make the shrimp extra-crunchy, add 1 tsp of caltrop starch and 1 tsp of salt to shelled and deveined shrimps. Mix well and leave them for 5 minutes. Then rinse and wipe dry them and cook as usual.

香山粉果 Xiangshan Fun Guo dumpling

皮料
Ingredients for dumpling skin
泰國薯粉 3 湯匙
澄麵半斤（2 杯）
生粉 2 湯匙
鹽半茶匙
豬油 1 湯匙
水 2 杯
3 tbsp Thai tapioca starch
300 g wheat starch (2 cups)
2 tbsp caltrop starch
1/2 tsp salt
1 tbsp lard
2 cups water

餡料 Filling
蝦肉半斤
甘筍 2 兩
瘦肉 4 兩
冬菇 3 朵
蝦米 1 兩
葱粒 2 湯匙
芫茜 2 棵
300 g shelled shrimps
75 g carrot
150 g lean pork
3 dried shiitake mushrooms
38 g dried shrimps
2 tbsp diced spring onion
2 sprigs coriander

蒸冬菇調味料
Seasoning for shiitake mushrooms
生抽、糖、油及紹酒各 1 茶匙
浸冬菇水 2 湯匙
1 tsp light soy sauce
1 tsp sugar
1 tsp oil
1 tsp Shaoxing wine
2 tbsp water in which shiitake
mushrooms are soaked

醃料 Marinade
鹽 1/4 茶匙
糖 1/4 茶匙
生抽 1/4 茶匙
生粉半茶匙
胡椒粉少許
1/4 tsp salt
1/4 tsp sugar
1/4 tsp light soy sauce
1/2 tsp caltrop starch
ground white pepper

調味料 Filling seasoning
鹽半茶匙
糖 3/4 茶匙
生抽 3/4 茶匙
麻油及胡椒粉各少許
1/2 tsp salt
3/4 tsp sugar
3/4 tsp light soy sauce
sesame oil
ground white pepper

外皮做法 Method for dumpling skin dough

1. 澄麵、薯粉及生粉篩勻，備用。
2. 煮滾水，下粉料攪勻，加入鹽及豬油煮溶，熄火，加蓋焗5分鐘。

1. Sieve wheat starch, tapioca starch and caltrop starch.

2. Boil water and add the dry ingredients. Mix well. Add salt and lard. Cook until lard melts. Turn off the heat. Cover the lid and leave it for 5 minutes.

餡料做法 Method for filling

1. 冬菇浸軟，去蒂，加入冬菇調味料蒸15分鐘，切粒。
2. 甘筍飛水，瀝乾水分，切粒；瘦肉切幼粒，下醃料拌勻；蝦米浸15分鐘，剁幼粒。
3. 蝦肉用生粉及鹽各半茶匙醃5分鐘，洗淨及抹乾備用。
4. 燒熱油2湯匙，先下瘦肉、冬菇及甘筍炒勻，加入葱粒拌炒，將材料移去鑊邊，下油1湯匙燒熱，加入蝦肉炒熱，再拌入蝦米，炒勻所有材料，潷酒，最後下調味料炒勻即可。

1. Soak shiitake mushrooms in water until soft. Cut off the stems. Add seasoning for shiitake mushrooms and mix well. Steam for 15 minutes. Finely dice them.

2. Blanch carrot in boiling water. Drain and dice it. Finely dice the lean pork and add marinade. Mix well. Soak dried shrimps in water for 15 minutes. Finely dice them.

3. Mix the shrimps with 1/2 tsp of caltrop starch and 1/2 tsp of salt. Mix well and leave them for 5 minutes. Rinse and wipe dry. Set aside.

4. Heat 2 tbsp of oil in a wok. Stir fry pork, shiitake mushrooms and carrot. Then add spring onion and stir further. Push the ingredients onto one side of the wok. Add 1 tbsp of oil and heat it up. Stir fry the shrimp until hot. Stir in dried shrimps. Toss all ingredients well. Lastly add filling seasoning. Mix well.

❧ 小煮意 ❧

- 皮料內加入薯粉，令外皮更挺身，外型美觀。

- 澄麵是麵粉去除蛋白質後的澱粉，沒黏性，焗煮後呈透明狀，多製成蝦餃或粉果等。

- Adding tapioca starch to the skin makes it firmer in texture so that the dumpling holds its shape better.

- Wheat starch is flour with its protein and gluten completely removed. It has no adhesiveness whatsoever and turns transparent after cooked. It is used to make Har Gow (shrimp dumpling) or Fun Guo dumpling skin.

包餡做法 Method for dumpling skin dough

1. 在桌面灑上薯粉，取出澄麵粉糰搓至幼滑，分成小粒狀，用菜刀或擀麵棒壓成窩形圓薄片。

2. 放入餡料 1 茶匙及芫茜葉，對摺包成粉果形，外皮塗抹少許油，放入蒸籠以大火蒸 5 分鐘即成。

1. Dust counter with tapioca starch. Knead the dumpling skin dough until smooth. Divide into small pieces. Roll each piece out into a thin round disc with its rim curling up.

3. Put 1 tsp of filling and a coriander leaf on the dough. Fold in half into a dumpling. Brush oil over it. Steam in a steamer for 5 minutes. Serve.

素粉果

少變化，多口味

Vegetarian Fun Guo dumpling

小煮意

若是素食者，可不加添芹菜及芫茜。

You may omit Chinese celery and coriander in this recipe if they work against your religion.

皮料
Ingredients for dumpling skin

泰國薯粉 3 湯匙
澄麵 2 杯
生粉 3 湯匙
鹽半茶匙
油 1 湯匙
水 2 杯或 2 1/2 杯

3 tbsp Thai tapioca starch
2 cups wheat starch
3 tbsp caltrop starch
1/2 tsp salt
1 tbsp oil
2 or 2 1/2 cups water

餡料 Filling

素火腿 1 條
冬筍 2 兩
甘筍 2 兩
冬菇 6 朵
馬蹄 6 粒
芹菜 3 棵
芫茜 2 棵（只取葉片）

1 piece vegetarian ham
75 g bamboo shoot
75 g carrot
6 dried shiitake mushrooms
6 water chestnuts
3 sprigs Chinese celery
2 sprigs coriander (leaves only)

蒸冬菇調味料
Seasoning for shiitake mushrooms

生抽、糖、油及紹酒各 1 茶匙
浸冬菇水 2 湯匙

1 tsp light soy sauce
1 tsp sugar
1 tsp oil
1 tsp Shaoxing wine
2 tbsp water in which
shiitake mushrooms are soaked

調味料
Filling seasoning

鹽半茶匙
糖半茶匙
生抽 2 茶匙
麻油及素蠔油各少許

1/2 tsp salt
1/2 tsp sugar
2 tsp light soy sauce
sesame oil
vegetarian oyster sauce

外皮做法 Method for dumpling skin dough

1. 澄麵、薯粉及生粉篩勻，備用。
2. 煮滾水，下鹽及油拌勻，再加入粉料攪勻，熄火，加蓋焗 5 分鐘。

1. Sieve wheat starch, tapioca starch and caltrop starch.
2. Boil water and add salt and oil. Put in the dry ingredients. Mix well. Turn off the heat. Cover the lid and leave it for 5 minutes.

餡料做法 Method for filling

1. 冬菇浸軟，去蒂，加入冬菇調味料蒸 15 分鐘，切絲。
2. 其餘材料切絲，備用。
3. 燒熱油 2 湯匙，下冬菇炒片刻，加入其他材料炒勻，拌入調味料，埋獻，拌勻即可上碟。

1. Soak shiitake mushrooms in water until soft. Cut off the stems. Add seasoning for shiitake mushrooms and mix well. Steam for 15 minutes. Finely shred them.
2. Shred remaining ingredients. Set aside.
3. Heat 2 tbsp of oil in a wok. Stir shiitake mushrooms briefly. Then add all remaining filling ingredients. Stir in filling seasoning. Thicken the caltrop starch slurry. Mix well.

包餡做法 Assembly

1. 在桌面灑上薯粉，取出澄麵粉糰搓至幼滑，分成小粒狀，用擀麵棒壓成窩形圓薄片。
2. 放入餡料 1 茶匙及芫茜葉，對摺包成粉果形，外皮塗抹少許油，以大火蒸 7 分鐘即成。

1. Dust counter with tapioca starch. Knead the dumpling skin dough until smooth. Divide into small pieces. Roll each piece out into a thin round disc with its rim curling up.
2. Put 1 tsp of filling and a coriander leaf on the dough. Fold in half into a dumpling. Brush oil over it. Steam in a steamer for 7 minutes. Serve.

Chaozhou Fun Guo dumpling

皮料
Ingredients for
dumpling skin

泰國薯粉 5 安士
（量杯計，約 60 克）
澄麵 8 安士
（量杯計，約 100 克）
生粉 2 湯匙
鹽半茶匙
豬油半湯匙
水 1 杯
60 g Thai tapioca starch
100 g wheat starch
2 tbsp caltrop starch
1/2 tsp salt
1/2 tbsp lard
1 cup water

餡料
Filling

蝦肉半斤
馬蹄 5 粒
芹菜 2 棵
瘦肉 4 兩
冬菇 5 朵
菜脯 5 湯匙
炸花生 5 湯匙
葱粒 2 湯匙
乾葱 2 粒（剁茸）
300 g shelled shrimps
5 water chestnuts
2 sprigs Chinese celery
150 g lean pork
5 dried shiitake mushrooms
5 tbsp salted radish
5 tbsp deep-fried peanuts
2 tbsp diced spring onion
2 shallots (finely chopped)

蒸冬菇調味料
Seasoning for shiitake mushrooms

生抽、糖、油及紹酒各 1 茶匙
浸冬菇水 2 湯匙
1 tsp light soy sauce
1 tsp sugar
1 tsp oil
1 tsp Shaoxing wine
2 tbsp water in which
shiitake mushrooms are soaked

醃料 Marinade

鹽 1/4 茶匙
糖 1/4 茶匙
生抽 1/4 茶匙
生粉半茶匙
胡椒粉少許
1/4 tsp salt
1/4 tsp sugar
1/4 tsp light soy sauce
1/2 tsp caltrop starch
ground white pepper

調味料 Seasoning

鹽半茶匙
糖 3/4 茶匙
生抽 1 茶匙
生粉、麻油及胡椒粉各少許
1/2 tsp salt
3/4 tsp sugar
1 tsp light soy sauce
caltrop starch
sesame oil
ground white pepper

外皮做法
Method for dumpling skin dough

1. 澄麵，薯粉及生粉篩勻，備用。
2. 煮滾水，加入鹽及豬油煮溶，下粉料攪勻，熄火，加蓋焗5分鐘。

1. Sieve wheat starch, tapioca starch and caltrop starch.

2. Boil water and add salt and lard. Cook until lard melts. Add the dry ingredients. Mix well. Turn off the heat. Cover the lid and leave it for 5 minutes.

 小煮意

製外皮時，澄麵放入滾水內拌勻，必須加蓋焗5分鐘才熟透。

When you make the dumpling skin dough, you must cover the lid and leave the dough in the pot for 5 minutes after you mix the wheat starch in boiling water. Otherwise, the dough isn't cooked through.

餡料做法 Method for filling

1. 冬菇浸軟，去蒂，加入冬菇調味料蒸 15 分鐘，切粒。
2. 蝦肉用生粉及鹽各半茶匙醃 5 分鐘，洗淨及抹乾備用。
3. 瘦肉切粒，下醃料拌勻；其餘材料切粒。
4. 燒熱油 2 湯匙，下乾蔥茸爆香，加入瘦肉、菜脯、蝦肉、冬菇、馬蹄及芹菜炒勻，灑入蔥粒及調味料拌炒，最後下炸花生即可。

1. Soak shiitake mushrooms in water until soft. Cut off the stems. Add seasoning for shiitake mushrooms and mix well. Steam for 15 minutes. Finely dice them.

2. Mix the shrimps with 1/2 tsp of caltrop starch and 1/2 tsp of salt. Mix well and leave them for 5 minutes. Rinse and wipe dry. Set aside.

3. Dice the pork and add marinade. Mix well. Finely dice the remaining ingredients.

4. Heat 2 tbsp of oil in a wok. Stir fry shallot until fragrant. Add pork, salted radish, shrimps, shiitake mushrooms, water chestnuts and Chinese celery. Then add spring onion and filling seasoning. Stir further. Sprinkle with deep-fried peanuts. Mix well.

包餡做法 Assembly

1. 在桌面灑上薯粉，取出澄麵粉糰搓至幼滑，分成小粒狀，用擀麵棒壓成窩形圓薄片。
2. 放入餡料 1 茶匙，對摺包成粉果形，外皮塗抹少許油，放入蒸籠以大火蒸 7 分鐘即成。

1. Dust counter with tapioca starch. Knead the dumpling skin dough until smooth. Divide into small pieces. Roll each piece out into a thin round disc with its rim curling up.

2. Put 1 tsp of filling on the dough. Fold in half into a dumpling. Brush oil over it. Steam in a steamer for 7 minutes. Serve.

Soup-filled dumpling 灌湯餃

皮料
Ingredients for dumpling skin dough

麵粉 10 安士
生粉 3 湯匙
蛋黃 2 個
水 7 至 8 湯匙
食用鹼水 1/4 茶匙

1 cup plain flour
3 tbsp caltrop starch
2 egg yolks
7 or 8 tbsp water
1/4 tbsp food-grade lye

餡料 Filling

瘦肉 4 兩
蝦肉 5 兩
雞膶 2 個
冬菇 5 朵
大菜 5 條
椰菜 4 片

150 g lean pork
190 g shelled shrimps
2 chicken livers
5 dried shiitake mushrooms
5 strands agar-agar
4 cabbage leaves

蒸冬菇調味料
Seasoning for shiitake mushrooms
生抽、糖、油及紹酒各 1 茶匙
浸冬菇水 2 湯匙
1 tsp light soy sauce
1 tsp sugar
1 tsp oil
1 tsp Shaoxing wine
2 tbsp water in which
shiitake mushrooms are soaked

調味料
Filling seasoning

鹽 1 茶匙
生粉 1 茶匙
糖、麻油及胡椒粉各少許
1 tsp salt
1 tsp caltrop starch
sugar
sesame oil
ground white pepper

上湯料 Stock

清雞湯 1 碗
水 3 杯
鹽半茶匙
韭黃 1 兩
1 bowl chicken stock
3 cups water
1/2 tsp salt
38 g yellow chives

蘸汁
Dipping sauce

大紅浙醋及薑絲各適量
red vinegar
shredded ginger

皮料做法 Method for dumpling skin dough

麵粉及生粉篩勻，盛於碗內，拌入水、鹼水及蛋黃搓勻至軟滑，蓋上濕布待約半小時。

Sieve flour and caltrop starch into a bowl. Stir in water, lye and egg yolks. Mix until lump-free. Cover with the damp towel and sit for 30 minutes.

❧ 小煮意 ❧

大菜煮溶及凝固後，容易包入皮
內，蒸熟後的灌湯餃帶有濃濃的湯
汁，口感豐腴。

Agar-agar is the trick to soupy filling
in dumplings. It's a lot easier to
handle after you congeal the stock
into solid. The dumpling will burst
with rich stock.

餡料做法 Method for filling

1. 大菜洗淨,用清雞湯半杯慢火煮溶,冷藏至凝固成大菜糕,使用時切粒。

2. 冬菇用調味料蒸15分鐘,切粒;瘦肉及雞膶洗淨,分別切粒。蝦肉用生粉及鹽各半茶匙醃5分鐘,洗淨及抹乾,切粒。

3. 燒熱鑊,下油3湯匙爆香所有餡料,放入調味料拌勻,潛酒,盛起。

4. 將麵糰切成小粒,用擀麵棒擀薄成3吋大之圓餅狀,包入餡料及大菜糕,揑實,接口處放在椰菜上,隔水蒸10分鐘,取出,蘸大紅浙醋及薑絲享用。

1. Rinse the agar-agar. Boil 1/2 cup of chicken stock over low heat in a pot. Put in agar-agar and cook until it dissolves. Let cool and refrigerate until set. Dice right before use.

2. Soak shiitake mushroom in water until soft. Remove the stems. Add seasoning for shiitake mushrooms. Mix well and steam for 15 minutes. Dice them. Rinse the pork and chicken liver. Dice them separately. Mix the shelled shrimps with 1/2 tsp of caltrop starch and 1/2 tsp of salt. Leave them for 5 minutes. Rinse and wipe dry. Dice them.

3. Heat a wok and add 3 tbsp of oil. Stir fry all filling ingredients until fragrant. Add filling seasoning and stir well. Sizzle with wine. Set aside.

4. Cut the dough into small pieces. Roll them out into round discs about 3 inches in diameter with a rolling pin. Put in filling and the diced agar-agar. Fold the skin to seal. Put it on a cabbage leaf with the seam facing down. Steam for 10 minutes. Serve with red vinegar and shredded ginger on the side.

珍珠雞 Mini glutinous rice dumpling in lotus leaf

材料 Ingredients

糯米 1 斤
梅頭瘦肉 2 兩
雞 8 件（取肉）
蝦仁 2 兩
冬菇 4 朵
叉燒 2 兩
冬筍 2 兩
鹹蛋黃 2 個
荷葉 4 塊
白洋布 1 塊

600 g glutinous rice
75 g pork shoulder butt
8 pieces chicken (de-boned)
75 g shelled shrimps
4 dried shiitake mushrooms
75 g barbecue pork
75 g bamboo shoot
2 salted egg yolks
4 lotus leaves
1 sheet white muslin cloth

調味料 Seasoning

鹽半茶匙
蠔油 1 湯匙
麻油、糖、生抽及胡椒粉各少許

1/2 tsp salt
1 tbsp oyster sauce
sesame oil
sugar
light soy sauce
ground white pepper

獻汁 Thickening glaze

上湯半杯
生粉 1 湯匙
1/2 cup stock
1 tbsp caltrop starch

❧ 小煮意 ❧

荷葉用熱水灼至軟身，包餡料時較容易處理；包裹時荷葉底部向內，令製成品更美觀。

Blanching the lotus leaves soften them so that can wrap around the filling easily. To make the dumplings look better, you may put the filling on the underside of the leaf (i.e. the side where the stem comes out).

做法 Method

1. 糯米洗淨，用水浸約 4 小時，放入筲箕隔去水分，以大量沸水沖淨糯米，瀝乾。

2. 蒸籠內鋪入白洋布，排上糯米，隔水蒸 30 分鐘，盛起，下少許鹽拌勻。

3. 梅頭瘦肉切粗粒；雞肉切件，與瘦肉混和，下鹽、糖、薑汁及生粉各少許醃勻。

4. 蝦仁去腸，加生粉及鹽各半茶匙醃 5 分鐘，洗淨，抹乾備用。

5. 冬菇浸軟，去蒂，加入浸冬菇水 2 湯匙、生抽、糖、油及紹酒各 1 茶匙拌勻蒸 15 分鐘，切條。

6. 叉燒切粗粒；冬筍飛水，吸乾水分，切條。

7. 燒紅鑊，下油 1 湯匙，潷酒，放入調味料煮滾，下叉燒、瘦肉、冬筍、蝦仁炒勻，埋獻，盛起待涼。

8. 荷葉放入熱水內略灼，洗淨，抹乾水分，每塊剪開一半，塗油，鋪上一層薄糯米飯，下已煮的餡料及汁料，排入雞件、冬菇及鹹蛋黃，蓋上一層糯米飯，將荷葉摺入包好，如是者完成所有材料，隔水蒸約 20 分鐘，趁熱食用。

1. Rinse the rice and soak it in water for 4 hours. Drain. Rinse the rice with a large amount of boiling water. Drain again.

2. Line a steamer with white muslin. Put the rice in and steam for 30 minutes. Set aside. Add salt and mix well.

3. Dice pork shoulder butt coarsely. Cut chicken into pieces. Mix pork with chicken. Add a little salt, sugar, ginger juice and caltrop starch. Mix well.

4. Devein the shrimps. Add 1/2 tsp of caltrop starch and 1/2 tsp of salt. Mix well and leave them for 5 minutes. Rinse and wipe dry. Set aside.

5. Soak shiitake mushrooms in water until soft. Cut off the stems. Add 2 tbsp of soaking water for mushrooms, 1 tsp of light soy sauce, 1 tsp of sugar, 1 tsp of oil and 1 tsp Shaoxing wine. Mix well and steam for 15 minutes. Cut into strips.

6. Dice barbecue pork coarsely. Blanch bamboo shoot in boiling water. Drain and wipe dry. Cut into strips.

7. Heat a wok and add 1 tbsp of oil. Sizzle with wine. Add seasoning and bring to the boil. Add barbecue pork, pork shoulder butt, bamboo shoot and shrimps. Mix well. Stir in thickening glaze. Set aside to let cool.

8. Blanch the lotus leaves in boiling water. Rinse well and wipe dry. Cut in half for each lotus leaf. Brush oil on them. Spread a layer of glutinous rice over them. Put in the filling and sauce. Then arrange chicken, shiitake mushrooms and a salted egg yolk. Top with another layer of glutinous rice. Fold the lotus leaves upward into a packet. Repeat this step with the rest of the ingredients. Steam for 20 minutes. Serve hot.

小籠包
Xiaolongbao
(Steamed mini pork buns)

材料 Ingredients

麵粉 12 安士（量杯計）
梅頭豬肉 4 兩
薑茸 2 湯匙
滾水 4 安士
椰菜半斤（切絲）
350 ml flour (measured in cup)
150 g pork shoulder butt
2 tbsp grated ginger
100 ml boiling water
300 g white cabbage (finely shredded)

調味料 Seasoning

生抽 1 茶匙
鹽 3/4 茶匙
糖 1/4 茶匙
生粉 2 茶匙
上湯 3 湯匙
麻油、油及胡椒粉各少許
1 tsp light soy sauce
3/4 tsp salt
1/4 tsp sugar
2 tsp caltrop starch
3 tbsp stock
sesame oil
oil
ground white pepper

蘸汁 Dipping sauce

鎮江香醋及薑絲各適量
red vinegar
shredded ginger

∾ 小煮意 ∾

• 搓好的小籠包放在椰菜絲上蒸，以免黏着底部，亦可用紅蘿蔔或牛油紙代替。

• 想入口豐腴，建議用豬皮熬成汁，並冷藏成凍糕，切粒後包入餡料內，但一般家庭製作較繁複，故省略。

• Putting the buns over a bed of shredded cabbage so that they won't stick to the steamer. You can also use sliced carrot or baking paper instead.

• If you want to serve a juicy xiaolongbao, I suggested to boil the pork skin soup first, then frozen into a parfait. Dice them and wrap into the filling. But this step maybe relatively complicated, it is omitted in the recipe.

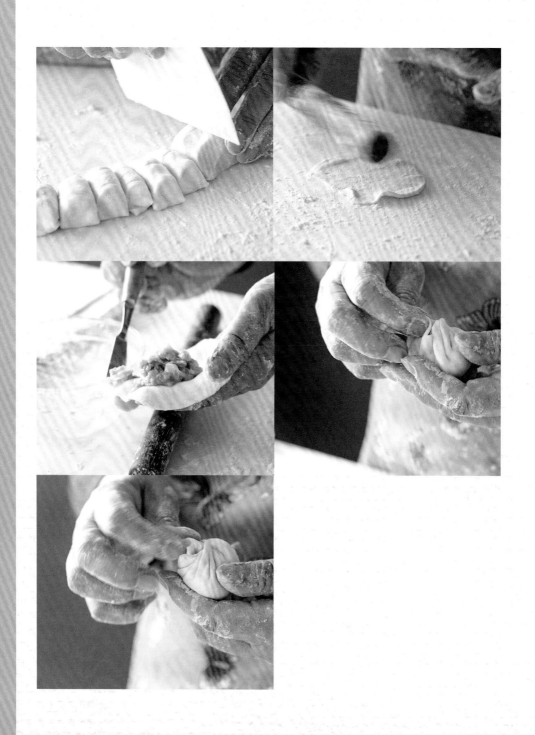

做法 Method

1. 麵粉篩勻，盛於碗內，慢慢下滾水拌勻，搓成幼滑粉糰，用濕布蓋着發酵15分鐘。

2. 豬肉剁碎，放入調味料拌勻，最後下薑茸攪勻。

3. 將麵糰搓成長條，切成小粒，用擀麵棒將麵糰擀成薄圓形狀，包入餡料，揑成小包狀，放在已鋪椰菜絲的蒸籠上，隔水大火蒸8分鐘，蘸鎮江香醋及薑絲食用。

1. Sieve flour into a bowl. Slowly stir in boiling water and work into smooth dough. Cover with damp towel and let it rest for 15 minutes.

2. Finely chop the pork. Add seasoning and mix well. Stir in grated ginger.

3. Roll the dough out into a long cylinder. Cut into small pieces. Roll each piece out into a round disc. Put some filling on it and fold the rim up into a small bun. Arrange the buns on a steamer lined with shredded cabbage. Steam over high heat for 8 minutes. Serve with red vinegar and shredded ginger on the side.

鮮蝦雲吞 Cantonese shrimp wantons

材料 Ingredients

鮮蝦肉半斤
梅頭豬肉半斤
韭黃 1 兩
清雞湯 5 碗
雲吞皮 4 兩

300 g shelled shrimps
300 g pork shoulder butt
38 g yellow chives
5 bowls chicken stock
150 g wanton skin

調味料 Seasoning

大地魚茸 2 茶匙
芝麻茸 1 茶匙
鹽 1 1/4 茶匙
糖 1 茶匙
生粉 1 湯匙
蛋黃 1 個
麻油及胡椒粉各少許

2 tsp ground dried plaice
1 tsp ground sesames
1 1/4 tsp salt
1 tsp sugar
1 tbsp caltrop starch
1 egg yolk
sesame oil
ground white pepper

做法 Method

1. 鮮蝦肉去腸,加生粉及鹽各半茶匙醃 5 分鐘,洗淨,抹乾。
2. 豬肉洗淨,剁茸;韭黃洗淨,切粒。
3. 將豬肉及蝦肉放碗內,加入調味料順一方向攪至起膠,放入雪櫃冷藏片刻。
4. 燒滾清雞湯,加入少許鹽、糖、麻油及胡椒粉,備用。
5. 餃皮內包入適量餡料,皮邊抹上水,用手略捏,放入滾水內煮熟至浮起,下清水半碗煮滾,盛於碗內。
6. 最後加入上湯,灑入韭黃即成。

1. Devein the shrimps. Add 1/2 tsp of caltrop starch and 1/2 tsp of salt. Mix well and leave them for 5 minutes. Rinse and wipe dry.

2. Rinse the pork and finely chop it. Rinse the yellow chives and dice them.

3. Put pork and shrimps into a bowl. Add seasoning and stir in one direction until sticky. Refrigerate briefly before use. This is the filling.

4. Boil the chicken stock. Add a pinch of salt, sugar, sesame oil and ground white pepper.

5. Put some filling on a piece of wanton skin. Wet the rim slightly. Fold to seal the seam. Boil in water until cooked. Pour in 1/2 bowl of water and bring to the boil. Transfer into a bowl.

6. Pour stock over the wantons. Sprinkle with yellow chives. Serve.

❧ 小煮意 ❧

大地魚又稱為方魚,用火焙乾,起肉去骨,打碎後即成大地魚茸。

Ground dried plaice, as its name suggests, is plaice that has been dried in open fire, de-boned and then ground into powder. It gives the stock the typical seafood sweetness you find in Cantonese wanton soup.

菜肉雲吞

少變化，多口味

材料 Ingredients

大白菜半斤
梅頭瘦肉 6 兩
蝦肉 4 兩
白方皮 12 兩
300 g Napa cabbage
225 g pork shoulder butt
150 g shelled shrimps
450 g white square wanton skin

上湯料 Stock

上湯 5 杯
鹽 1 茶匙
生抽 2 茶匙
麻油 1 茶匙
5 cups stock
1 tsp salt
2 tsp light soy sauce
1 tsp sesame oil

調味料 Seasoning

鹽 1 茶匙
糖半茶匙
雞粉半茶匙
生抽 1 茶匙
麻油及胡椒粉各少許
麵粉 2 湯匙
蛋液 2 湯匙
1 tsp salt
1/2 tsp sugar
1/2 tsp chicken bouillon powder
1 tsp light soy sauce
sesame oil
ground white pepper
2 tbsp flour
2 tbsp whisked egg

Cabbage and pork wantons

做法 Method

1. 大白菜洗淨，飛水至軟身，沖凍水，擠乾水分，切幼粒。

2. 豬肉洗淨，抹乾水分，剁茸；蝦肉去腸，用生粉及鹽各半茶匙醃5分鐘，洗淨，抹乾水分，切粒。

3. 白菜粒、肉茸及及蝦肉粒放入碗內，拌入調味料，順一方向攪至起膠，在碗內來回撻數次，放入雪櫃冷藏15分鐘。

4. 白方皮內包入餡料1湯匙，皮邊抹上生粉水，用手略捏。

5. 燒滾一鍋水，放入菜肉雲吞，煮滾後再倒入半碗水，待滾一會，盛於碗內。

6. 煮滾上湯，倒入碗內即可品嘗。

1. Rinse the cabbage. Blanch in boiling water until soft. Rinse in cold water. Squeeze dry. Finely dice it.

2. Rinse the pork and wipe dry. Finely chop it. Devein the shrimps. Add 1/2 tsp of caltrop starch and 1/2 tsp of salt. Mix well and leave for 5 minutes. Rinse and wipe dry. Dice them.

3. Put cabbage, chopped pork and diced shrimps into a bowl. Add seasoning and stir in one direction until sticky. Lift the mixture up and slap it hard into the bowl a few times. Refrigerate for 15 minutes. This is the filling.

4. Put 1 tbsp of filling on the wanton skin. Wet the rim with caltrop starch slurry. Pinch the corners randomly together.

5. Boil a pot of water. Put in the wantons. Boil until they float. Put in 1/2 bowl of cold water and bring to the boil again. Transfer the wantons into a serving bowl.

6. Boil the stock and pour over the wantons. Serve.

香茜餃 Steamed coriander dumpling

材料 Ingredients

圓型白色餃皮半斤
蝦仁 6 兩
半肥瘦豬肉 5 兩
冬菇 5 朵
芫茜 3 棵
蔥白茸 3 湯匙
300 g round white dumpling skin
225 g shelled shrimps
190 g half-fatty pork
5 shiitake mushrooms
3 sprigs coriander
3 tbsp finely chopped spring onion
(white part only)

蒸冬菇調味料
Seasoning for shiitake mushrooms

生抽、糖、油及紹酒各 1 茶匙
浸冬菇水 2 湯匙
1 tsp light soy sauce
1 tsp sugar
1 tsp oil
1 tsp Shaoxing wine
2 tbsp water in which
shiitake mushrooms are soaked

調味料 Seasoning

鹽半茶匙
糖 1/4 茶匙
生抽 1 茶匙
生粉 1 湯匙
上湯 4 湯匙
麻油及胡椒粉各少許
1/2 tsp salt
1/4 tsp sugar
1 tsp light soy sauce
1 tbsp caltrop starch
4 tbsp stock
sesame oil
ground white pepper

做法 Method

1. 冬菇浸軟，去蒂，加入冬菇調味料蒸 15 分鐘，切粒。
2. 蝦肉去腸，加生粉及鹽各半茶匙醃 5 分鐘，洗淨及抹乾，切粒。
3. 瘦肉剁成茸；芫茜切碎。
4. 將肉茸、蝦仁、冬菇、蔥白茸及調味料拌勻，灑入芫茜碎略拌，冷藏備用。
5. 將適量餡料放於餃皮內，對摺，以生粉水收口，揑實，放入已塗油的蒸籠內，隔水蒸 5 分鐘，即可。

1. Soak shiitake mushrooms in water until soft. Cut off the stems. Add seasoning for shiitake mushrooms and mix well. Steam for 15 minutes. Finely dice them.
2. Devein the shrimps. Add 1/2 tsp of caltrop starch and 1/2 tsp of salt. Mix well and leave them for 5 minutes. Rinse and wipe dry. Dice them.
3. Finely chop the pork and the coriander separately.
4. In a mixing bowl, put in chopped pork, shrimps, shiitake mushrooms, finely chopped spring onion and filling seasoning. Mix well. Sprinkle with coriander. Toss again. Refrigerate for later use. This is the filling.
5. Put some filling on the dumpling skin. Brush some caltrop starch slurry on the rim. Fold in half and press the rim firmly. Put into a grease steamer. Steam for 5 minutes. Serve.

❧ 小煮意 ❧

餃子蒸熟後，也可用油煎至兩面金黃色及香脆享用，有不同的食味效果。
After steaming the dumplings, you may fry them until both sides golden and serve. The dumplings taste completely different.

眉豆茶粿

Hakka dumpling with black-eyed bean filling

材料 Ingredients

眉豆 12 兩
葱 5 棵（切段）
五香粉 4 茶匙
糯米粉 1 斤
粘米粉 4 安士（量杯計）
豬油 1 湯匙
糉葉或蕉葉數片
450 g black-eyed beans
5 sprigs spring onion
 (cut into short lengths)
4 tsp five-spice powder
600 g glutinous rice flour
115 ml long-grain rice flour
(measured in cup)
1 tbsp lard
bamboo or banana leaves

餡料做法 Method for filling

1. 眉豆用水浸約 1 小時，去掉豆殼，放入滾水內煲半小時，熄火，焗片刻，隔乾水分。

2. 燒熱鑊，下油 5 湯匙，放入葱段爆香，棄去；下眉豆並用鑊鏟壓扁豆粒，灑入五香粉及鹽少許，炒至水分收乾，即成眉豆蓉。

3. 糉葉洗淨，放入滾水內，下少許油略煮，修剪整齊。

1. Soak black-eyed beans in water for 1 hour. Remove the shells. Boil in water for 30 minutes. Turn off the heat and cover the lid. Leave them briefly. Strain.

2. Heat a wok and add 5 tbsp of oil. Stir fry spring onion until fragrant. Discard the spring onion. Put the black-eyed beans into the fragrant oil and mash them with a spatula. Sprinkle with five-spice powder and a pinch of salt. Stir until it dries up. This is the filling.

3. Rinse the bamboo leaves. Put them in boiling water and add a little oil. Cook briefly and drain. Trim them neatly.

小煮意

要密切留意蒸茶粿的時間，不可逾時。

Make sure you steam the dumplings according to the instruction. Do not overcook them.

粉糰做法 Dough and assembly

1. 取糯米粉 18 安士，用清水 9 安士搓勻。

2. 燒滾水 4 杯，將以上糯米粉糰分數次加入，煮片刻至半熟，瀝乾水分，保持和暖。

3. 餘下的糯米粉、粘米粉及鹽半茶匙拌勻，加入半熟粉糰及豬油搓和，分切成小粒，包入眉豆蓉，放在塗抹少許油的欓葉上，蒸約 8 分鐘即可。

1. Mix 530 ml of glutinous rice flour with 225 ml of water. Knead well.

2. Boil 4 cups of water. Divide the dough into a few pieces. Put them in the water and cook until half-done. Drain. Keep them warm.

3. In a mixing bowl, put in the remaining glutinous rice flour, long-grain rice flour and 1/2 tsp of salt. Mix well. Put in the half-cooked dough and lard. Knead to mix well. Divide into small pieces. Stuff the dough with the filling. Put it on a piece of greased bamboo leaf. Steam for 8 minutes. Serve.

Pea sprout and
scallop dumpling 豆苗帶子餃

皮料
Ingredients for dumpling skin

澄麵半斤（約 12 安士）
泰國薯粉 1 湯匙
生粉 2 湯匙
鹽半茶匙
豬油 1 湯匙
水 2 1/4 杯

300 g wheat starch
1 tbsp Thai tapioca starch
2 tbsp caltrop starch
1/2 tsp salt
1 tbsp lard
2 1/4 cups water

餡料 Filling

鮮蝦肉 4 兩
帶子半磅
肥豬肉 1 湯匙
豆苗 4 兩
蛋白 1 個（後下）

150 g shelled shrimps
225 g scallops
1 tbsp fatty pork
150 g pea sprouts
1 egg white (added last)

調味料 Seasoning

鹽 3/4 茶匙
糖半茶匙
生粉 2 湯匙
蛋白 1 個
紹酒、麻油及胡椒粉各少許

3/4 tsp salt
1/2 tsp sugar
2 tbsp caltrop starch
1 egg white
Shaoxing wine
sesame oil
ground white pepper

外皮做法
Method for dumpling skin dough

1. 澄麵、生粉、薯粉及鹽篩勻。

2. 煮滾水，下豬油略煮，加入以上粉料，用木棒迅速攪拌，熄火，加蓋焗 5 分鐘。

1. Sieve wheat starch, caltrop starch, tapioca starch and salt.

2. Boil water and add the lard until melts. Add the dry ingredients and stir well. Turn off the heat. Cover the lid and leave it for 5 minutes.

餡料做法 Method for filling

1. 蝦肉去腸，加生粉及鹽各半茶匙醃5分鐘，洗淨，抹乾備用。
2. 帶子飛水，抹乾，切成4份；肥豬肉煮熟，切幼粒。
3. 燒滾水，下鹽、糖、紹酒及油各少許，放入豆苗略灼，擠乾水分，切絲。
4. 所有材料放於大碗內，下調味料拌勻，冷藏備用。

1. Devein the shrimps. Add 1/2 tsp of caltrop starch and 1/2 tsp of salt. Mix well and leave them for 5 minutes. Rinse and wipe dry. Set aside.

2. Blanch the scallops in boiling water. Wipe dry. Cut each into 4 pieces. Boil the fatty pork in water until done. Dice finely.

3. Boil water. Put in a pinch of salt, sugar, Shaoxing wine and oil. Blanch the pea sprouts briefly. Drain and squeeze dry. Finely shred them.

4. Put all ingredients into a mixing bowl. Add seasoning and stir well. Refrigerate for later use.

包餡做法 Assembly

1. 在桌面灑上薯粉，取出粉糰搓至幼滑，分成小粒狀，用擀麵棒壓成窩形圓薄片。
2. 沿邊摺成餃形，釀入適量餡料，將帶子鋪面，以杞子裝飾，放入已塗油蒸籠內，以大火蒸7分鐘即成。

1. Dust counter with tapioca starch. Knead the dumpling skin dough until smooth. Divide into small pieces. Roll each piece out into a thin round disc with its rim curling up.

2. Pleat the rim of the dumpling skin closer to you so it looks like a bowl. Put some filling on the dough. Place the scallop on the top, then decorate with Qi Zi. Steam in a greased steamer for 7 minutes. Serve.

∽ 小煮意 ∽

豆苗先用鹽、糖及紹酒飛水，可辟除豆腥味。

Blanching pea sprouts in salt, sugar and Shaoxing wine helps remove their grassy taste.

叉燒包（傳統版）

Char-siu bao
(Steamed barbecue pork buns)
Traditional version

以老麵糰發酵製作包子，步驟雖然有點繁複，但包子出來的效果卻與別不同，有興趣的不妨試試看！

This recipe involves pre-ferment. The steps could be a bit intimidating, but the buns have a special texture not achievable otherwise. Give it a try.

老麵糰材料 Pre-ferment
低筋麵粉 90 克（篩勻）
乾酵母 1/4 茶匙
暖水 45 克（約 20℃）
90 g cake flour (sieved)
1/4 tsp dry yeast
45 g warm water at 20°C

新麵糰材料 Dough
低筋麵粉 125 克（篩勻）
乾酵母 1 1/2 茶匙
泡打粉 1 1/2 茶匙
砂糖 50 克
白油 1 湯匙
暖水 40 克（約 20℃）
125 g cake flour (sieved)
1 1/2 tsp dry yeast
1 1/2 tsp baking powder
50 g sugar
1 tbsp shortening
40 g warm water at 20°C

餡料 Filling
叉燒 6 兩（切粒）
葱 2 棵（切短度）
乾葱 2 粒（拍碎）
225 g barbecue pork (diced)
2 sprigs spring onion
 (cut into short lengths)
2 shallots (crushed)

調味料 Seasoning
鹽及胡椒粉各 1/4 茶匙
生抽 1 茶匙
老抽 1 茶匙
蠔油 1 1/2 茶匙
麻油少許
水 2 安士
黃糖碎 2 湯匙
1/4 tsp salt
1/4 tsp ground white pepper
1 tsp light soy sauce
1 tsp dark soy sauce
1 1/2 tsp oyster sauce
sesame oil
60 ml water
2 tbsp brown sugar slab (chopped)

獻汁 Thickening glaze
生粉 2 湯匙
水 1/4 杯
2 tbsp caltrop starch
1/4 cup water

老麵糰做法 Method for pre-ferment

1. 乾酵母及暖水拌勻，加入低筋麵粉混合，用手搓揉至均勻成光滑麵糰。

2. 蓋上濕布或保鮮紙，待發酵漲大至兩倍（夏季約需 2 小時；冬季約 3 小時）。

1. Mix dry yeast with warm water and mix well. Add cake flour and stir well. Knead into smooth dough with your hands.

2. Cover with damp towel or cling film and let the dough ferment until it doubles in size (for about 2 hours in summer or 3 hours in winter).

新麵糰做法 Method for dough

1. 低筋麵粉放於桌面，中間開一穴，築成粉牆。

2. 在中間放入砂糖、白油、乾酵母及泡打粉，用手慢慢拌勻至糖溶化，逐少加入暖水搓勻，放入老麵糰，徐徐拌入麵粉，用手搓至滑身成新麵糰，蓋上濕布待 20 分鐘發酵。

1. Pour the cake flour on a counter. Make a well at the centre.

2. Put sugar, shortening, dry yeast and baking powder in the well. Rub these ingredients with your fingers until sugar dissolves. Slowly add warm water and mix well. Put in the pre-ferment. Push the cake flour toward the well and mix well. Knead into smooth dough. Cover the dough in damp towel and leave it to proof for 20 minutes.

餡料做法 Method for filling

燒熱油3湯匙，下葱段及乾葱爆香，棄去，加入叉
燒炒勻，拌入調味料，最後埋獻，待涼，冷藏更佳。

Heat 3 tbsp of oil in a wok. Stir fry spring onion
and shallots until fragrant. Discard them. Put
in the barbecue pork and stir fry briefly. Stir in
seasoning and then thickening glaze. Let cool. (It
works even better if refrigerated.)

綜合做法 Assembly

麵糰分切成小粒，擀薄，包入叉燒餡料，收緊邊緣，
揑成包子狀，貼在牛油紙上，放於蒸籠用大火隔水
蒸15分鐘即成。

Cut the dough into small pieces. Roll out each
piece of dough. Wrap some filling with the dough.
Seal the seam and shape into a bun. Put it on a
baking paper and then into a steamer. Steam over
high heat for 15 minutes. Serve.

❦ 小煮意 ❧

- 搓揉新麵糰時，要待砂糖完全溶化，才拌入暖水。

- 加入的暖水不宜太熱，微暖即可，否則搓出來的麵糰容易起筋。

- 老、新麵糰搓和後，發酵 20 分鐘後需立即包餡及蒸熟，否則麵糰容易變軟，包子定型不足。

- When you knead the dough, make sure the sugar has dissolved before stirring in warm water.

- The warm water should not be hot. It should just be lukewarm. Otherwise, the dough may turn rubbery.

- After you mix the pre-ferment with the main dough, you should only let it proof for only 20 minutes before you wrap in the filling and steam them. Over-fermented dough tends to make the buns too soft and they can't hold their shape well.

這個製叉燒包的簡易版本，是我自己設計及研發的，希望大家在家可以輕鬆製作鬆軟的叉燒包。

I designed and invented this recipe myself. I hope everyone can make fluffy Char-siu bao at home.

小煮意

搓皮時加入白醋，令包皮白晢，達天然漂白的功效，而且亦較健康。

The white vinegar keeps the bun snow white. It is a natural bleach that is good for health.

材料 Ingredients

麵粉 1 斤（32 安士，約 4 量杯）
發粉 6 茶匙
砂糖 6 兩（8 安士）
豬油 2 湯匙
清水 6 安士至 8 安士
白醋 1 茶匙
牛油紙 2 1/2 吋丁方塊
600 g plain flour
6 tsp self-raising flour
225 g sugar
2 tbsp lard
175-236 ml water
1 tsp white vinegar
2 1/2 inch square baking paper

餡料 Filling

叉燒 8 兩（切粒）
葱 2 棵（切短度）
乾葱 2 粒（拍碎）
300 g barbecue pork (diced)
2 sprigs spring onion (cut into short lengths)
2 shallots (crushed)

調味料 Seasoning

鹽及胡椒粉各 1/4 茶匙
生抽 1 茶匙
老抽及蠔油各 2 茶匙
麻油 1 茶匙
水半杯
黃糖 3 湯匙（切碎）
1/4 tsp salt
1/4 tsp ground white pepper
1 tsp light soy sauce
2 tsp dark soy sauce
2 tsp oyster sauce
1 tsp sesame oil
1/2 cup water
3 tbsp brown sugar slab (chopped)

獻汁 Thickening glaze

生粉 3 湯匙
水 1/4 杯
3 tbsp caltrop starch
1/4 cup water

Char-siu bao
(Steamed barbecue pork buns)
Easy version

做法 Method

1. 燒熱油 3 湯匙，下蔥段及乾蔥爆香，棄去，加入叉燒炒勻，拌入調味料，最後埋獻拌和。

2. 麵粉及發粉混和，篩勻後放於桌面。於麵粉中間開一穴，加入砂糖及豬油，用手慢慢搓勻至糖溶化，徐徐加入水及白醋，拌入麵粉輕手搓至滑，用濕布蓋着發酵 30 分鐘，分切成小粒。

3. 擀薄包皮，包入叉燒餡料，收緊邊緣，包成包子狀，貼在牛油紙上，放入蒸籠以大火蒸 15 分鐘即可。

1. Heat 3 tbsp of oil in a wok. Stir fry shallots and spring onion until fragrant. Discard them. Put in the barbecue pork and stir fry briefly. Add seasoning. Stir in thickening glaze and mix well. This is the filling.

2. Sieve flour and self-raising flour together on a counter. Make a well at the centre. Put in sugar and lard. Rub the sugar and lard with your fingers until the sugar dissolves. Slowly add water and white vinegar. Push the flour into the well and knead into smooth dough. Cover with damp towel and let it rest for 30 minutes. Cut the dough into small pieces.

3. Roll out each piece of dough. Wrap some filling with the dough. Seal the seam and shape into a bun. Put it on a baking paper and then into a steamer. Steam over high heat for 15 minutes. Serve.

日式窩貼

Gyoza
(Japanese pot-stickers)

皮料
Ingredients for
dumpling skin dough

麵粉 16 安士（量杯計）
高筋麵粉 1 湯匙
鹽 3/4 茶匙
滾水 7 安士

470 ml plain flour (measured in cup)
1 tbsp bread flour
3/4 tsp salt
200 ml boiling water

餡料 Filling

梅頭豬肉半斤
薑茸 2 湯匙
葱白 2 條（剁茸）

300 g pork shoulder butt
2 tbsp grated ginger
2 sprig spring onion (white part
only, finely chopped)

調味料 Seasoning

生抽 1 茶匙
鹽 1 茶匙
生粉 1 湯匙
上湯 3 湯匙
蛋液半個
薑茸 2 湯匙
葱白粒 3 湯匙
麻油、糖及胡椒粉各少許

1 tsp light soy sauce
1 tsp salt
1 tbsp caltrop starch
3 tbsp stock
1/2 egg (whisked)
2 tbsp grated ginger
3 tbsp diced spring onion
 (white part only)
sesame oil
sugar
ground white pepper

做法 Method

1. 麵粉、高筋麵粉及鹽篩勻，盛於碗內，加入滾水用膠刮刀迅速拌勻，搓成麵糰，用濕布蓋着待半小時。

2. 豬肉剁碎，放入薑茸、葱白茸及調味料拌勻。

3. 將麵糰搓成光滑表面，切成棋子般大小，用擀麵棒擀成塊狀，包入餡料 1 湯匙，捏成餃子狀。

4. 燒熱鑊，下少許油，排上鍋貼以中火煎一會，加入水半碗，加蓋，煮至油分作響，揭開鑊蓋，轉大火再煎一會即可。

1. Sieve plain flour, bread flour and salt into a bowl. Add boiling water and stir quickly with a spatula. Knead into dough. Cover with damp towel for 30 minutes.

2. Finely chop the pork. Add ginger, spring onion and seasoning. Mix well.

3. Knead the dough until smooth. Cut into pieces about the size of a chess. Roll each out into a thin disc. Put 1 tbsp of filling on it. Fold in half and pleat the seam.

4. Heat oil in a pan and arrange the dumplings neatly on it. Fry over medium heat for a while. Then add a half bowl of water. Cover the lid and cook until the water dries out and you hear the oil sizzles. Open the lid and turn to high heat. Keep frying briefly to brown them. Serve.

～ 小煮意 ～

- 煎鍋貼時加水蓋着，能夠將鍋貼煎得全熟，但緊記只煎底部。

- 加蓋後，聽到油脂嘓嚦啪喇作響，即代表水分收乾，皮脆香口。

- When you fry the gyoza, adding water and covering the lid help cook them through. Make sure you only fry one side of them.

- After you add water and cover the lid, sizzling sound means the water dries out and the oil is frying the dumpling skin for a crispy mouthfeel.

鍋貼

皮料
Ingredients for dumpling skin dough

麵粉半斤（2 1/4 杯）
鹽半茶匙
凍水 3 安士
滾水 4 安士

300 g (2 1/4 cups) plain flour
1/2 tsp salt
90 ml cold water
125 ml boiling water

餡料 Filling

梅頭豬肉半斤（剁碎）
馬蹄 4 粒（去皮、拍碎）

300 g pork shoulder butt (finely chopped)
4 water chestnuts (peeled and crushed)

調味料 Seasoning

鹽 1 茶匙
糖半茶匙
粟粉 1 湯匙
上湯 3 湯匙
麻油及胡椒粉各少許

1 tsp salt
1/2 tsp sugar
1 tbsp cornstarch
3 tbsp stock
sesame oil
ground white pepper

Shanghainese pot stickers

做法 Method

1. 麵粉篩勻，分成兩等份；取一份用凍水搓勻，另一份與沸水搓勻，再將兩份麵糰拌和，下鹽搓成軟麵糰，用濕布蓋着待10分鐘。
2. 豬肉茸、馬蹄碎及調味料拌勻。
3. 將麵糰搓成光滑表面，切成棋子般大小，用擀麵棒擀成塊狀，包入餡料1湯匙，捏實。
4. 燒熱鑊，下少許油，排入鍋貼以中火煎2分鐘，加入水半杯，加蓋，以小火再煎5分鐘，揭開鑊蓋，再下少許油煎3分鐘，至底部呈金黃色，反扣上碟即成。

1. Sieve plain flour and divide into two equal portions. Add cold water to one portion and boiling water to the other. Knead separately until smooth. Then mix them together and knead until well incorporated. Add salt and knead into soft dough. Cover with damp towel for 10 minutes.

2. Mix together the pork, crushed water chestnuts and seasoning.

3. Knead the dough until smooth. Cut into pieces about the size of a chess. Roll each out into a thin disc. Put 1 tbsp of filling on it. Fold in half and pleat the seam.

4. Heat oil in a pan and arrange the dumplings neatly on it. Fry over medium heat for 2 minutes. Then add 1/2 cup of water. Cover the lid and cook over low heat for 5 minutes. Open the lid and add oil. Fry for 3 more minutes until one side golden. Turn the pot stickers out on a plate. Serve.

小煮意

- 調味料加了上湯，令餡料充滿湯汁的精華。

- 將麵粉分成兩份，分別以凍水及沸水搓勻後混和，目的是令麵糰軟硬適中，若只用凍水會太硬；用沸水搓則太軟。

- Adding stock to the seasoning keeps the filling moist and full of meaty flavour.

- I divided the flour into two parts and add hot and cold water to each part before mix them together. This step helps achieve the optimal consistency of the dough. It tends to be too hard if only cold water is added. On the other hand, it tends to be too soft if only hot water is added.

粢
飯　Shanghainese sticky rice roll

材料 (可製成 4 個)
Ingredients
(makes 4 rolls)

糯米 1 1/2 斤
榨菜 3 兩
豬肉鬆 3 兩
油條 2 條
白洋布 2 張

900 g glutinous rice
113 g Zha Cai (spicy preserved mustard tuber)
113 g pork floss
2 deep-fried dough sticks
2 sheets white muslin cloth

> **✎ 小煮意 ✎**
>
> 糯米用滾水略沖，可去掉表面膠質。
>
> Rinsing the glutinous rice with boiling water helps remove part of its sticky substance.

做法 Method

1. 糯米洗淨，用水浸 6 小時，瀝乾水分，用滾水沖淨，放入鹽及油各 1 茶匙拌勻。

2. 白洋布放於蒸籠，排上糯米隔水蒸 30 分鐘。

3. 榨菜洗淨，擠乾水分，用糖及麻油各 1 茶匙拌勻。

4. 將每條油條撕成兩小條，再切半。

5. 將濕布擠乾，鋪於桌面上，抹上一層糯米，放上油條 1/4 條、肉鬆及榨菜適量，再鋪上糯米，將布扭緊即可。

1. Rinse the rice and soak it in water for 6 hours. Drain and rinse with boiling water. Drain again. Add 1 tsp of salt and 1 tsp of oil. Mix well.

2. Line a steamer with white muslin. Put in the rice and steam for 30 minutes.

3. Rinse the Zha Cai and squeeze dry. Add 1 tsp of sugar and 1 tsp of sesame oil. Mix well.

4. Divide each deep-fried dough stick into two along the length. Then cut each half into quarters.

5. Wet another sheet of muslin cloth and wring it out. Lay flat on the counter. Spread a layer of glutinous rice on the cloth. Put on a quarter of a deep-fried dough stick, pork floss and some Zha Cai. Spread some more rice on top. Roll the cloth up and twist both ends tightly. Remove the cloth and serve.

鮮蝦餃

Har Gow
（Cantonese shrimp dumpling）

皮料
Ingredients for dumpling
skin dough

澄麵半斤（約 12 安士）
泰國薯粉 2 湯匙
生粉 2 湯匙
鹽半茶匙
豬油 1 湯匙
水 2 杯

300 g wheat starch
2 tbsp Thai tapioca starch
2 tbsp caltrop starch
1/2 tsp salt
1 tbsp lard
2 cups water

調味料 Seasoning

鹽半茶匙
糖半茶匙
麻油半茶匙
胡椒粉 1/4 茶匙
生粉 1 湯匙

1/2 tsp salt
1/2 tsp sugar
1/2 tsp sesame oil
1/4 tsp ground white pepper
1 tbsp caltrop starch

餡料 Filling

鮮蝦 1 斤
冬筍 10 兩
肥豬肉 2 兩

600 g shrimps
375 g bamboo shoot
75 g fatty pork

小煮意

- 將冬筍飛水，切成絲，用毛巾擠乾水分，可去除冬筍的異味。

- 除用擀木棒外，一般家庭可利用菜刀面，在小粒粉糰上旋轉並按壓，可製成又薄又圓的蝦餃皮。

- To remove the unappetizing taste in bamboo shoot, blanch it and shred it first. Then wrap it in a towel and squeeze dry.

- Apart from using a rolling pin, you may also use a cleaver to flatten the dough. Just press the dough with the flat side of the knife from the centre toward the edge once. Turn rotate it slightly and press again. You'd end up with a thin and round skin when you do it repeatedly.

Video

外皮做法 Method for dumpling skin dough

1. 澄麵、生粉及鹽篩勻，備用。
2. 煮滾水，下豬油略煮，加入粉料用木棒拌勻，熄火，加蓋焗 5 分鐘。
3. 桌面灑上薯粉，取出粉糰搓至幼滑，用濕布覆蓋待 10 分鐘。

1. Sieve wheat starch, caltrop starch and salt.
2. Boil water and add lard to cook briefly. Add dry ingredients and mix well with a wooden spatula. Turn off the heat. Cover the lid and leave it for 5 minutes.
3. Dust counter with tapioca starch. Knead the dumpling skin dough until smooth. Cover with a damp towel and let it rest for 10 minutes.

餡料做法 Method for filling

1. 鮮蝦肉去腸，加生粉及鹽各半茶匙醃 5 分鐘，洗淨及抹乾，切半。
2. 冬筍洗淨，切幼粒；肥豬肉焓熟，切幼粒。
3. 餡料與調味料拌勻，備用。

1. Devein the shrimps. Add 1/2 tsp of caltrop starch and 1/2 tsp of salt. Mix well and leave them for 5 minutes. Rinse and wipe dry. Divide into 2 equal portions.
2. Rinse bamboo shoot and dice finely. Boil the fatty pork in water until done. Dice finely.
3. Put all ingredients into a mixing bowl. Add seasoning and stir well.

包餡做法 Assembly

1. 麵糰搓成條狀，切小粒，用擀麵棒擀薄，壓成窩形圓薄片。

2. 先將後方的皮打摺，捏成窩形狀，包入適量餡料，再將另一邊的皮覆上，推捏成蝦餃形狀。

3. 排入已塗油蒸籠內，隔水以大火蒸 7 分鐘即可。

1. Dust counter with tapioca starch. Roll the dough into a cylinder. Divide into small pieces. Roll each piece out into a thin round disc with its rim curling up.

2. Pleat the rim of the dumpling skin closer to you so that it looks like a bowl. Put in some filling. Fold the far end of the dumpling skin toward you. Pinch and press firmly into a typical crescent shape.

3. Arrange the dumplings in a greased steamer. Steam for 7 minutes. Serve.

少變化，多口味

雲耳鮮蝦餃

材料
Ingredients

鮮蝦仁 6 兩
梅頭瘦肉 6 兩
雲耳 3 錢
芹菜 4 兩
白色圓形餃皮 1 包
225 g shelled shrimps
225 g pork shoulder butt
12 g cloud ear fungus
150 g Chinese celery
1 pack round white dumpling skin

調味料 Seasoning

鹽 1 茶匙
生抽 2 茶匙
蛋黃 1 個
糖及麻油各少許
1 tsp salt
2 tsp light soy sauce
1 egg yolk
sugar
sesame oil

Steamed cloud ear shrimp dumpling

瘦肉用刀細剁，比用攪肉機製成的肉茸，口感更富彈性。製作餃子餡料，我喜歡用刀剁肉的口感。

Finely chopped pork has a more springy texture than that ground with a machine. I prefer chopped pork in my dumpling filling for its mouthfeel.

做法 Method

1. 鮮蝦仁去腸，加生粉及鹽各半茶匙醃 5 分鐘，洗淨及抹乾，切粗粒。
2. 雲耳浸軟，洗淨，切條；芹菜切去根部，去葉，洗淨，切粒；瘦肉洗淨，切粒，剁成肉茸。
3. 全部材料與調味料攪拌均勻，備用。
4. 餃皮內包入適量餡料，對摺，用水封口，排入碟內，灑少許水，隔水蒸約 8 分鐘即可。

1. Devein the shrimps. Add 1/2 tsp of caltrop starch and 1/2 tsp of salt. Mix well and leave them for 5 minutes. Rinse and wipe dry. Dice them coarsely.

2. Soak cloud ear in water until soft. Rinse and cut into strips. Cut off the roots of the Chinese celery. Discard the leaves and use the stems only. Rinse and dice them. Rinse the pork. Dice finely. Then chop finely.

3. Mix all ingredients and seasoning together.

4. Put some filling onto each piece of dumpling skin. Wet the rim with water. Fold in half and seal well. Arrange on a plate. Spray some water on them. Steam for 8 minutes. Serve.

少變化，多口味

菠菜皮鮮蝦餃

❧ 小煮意 ❧

餡料拌入調味料後，放進雪櫃冷藏，令蝦肉更爽口。

After you stir in the seasoning into the filling, it's advisable to refrigerate it briefly. That would make the shrimp crunchy and firmer in texture.

菠菜皮料
Ingredients for dumpling skin dough

菠菜半斤
澄麵半斤（約 12 安士）
泰國薯粉 1 湯匙
生粉 2 湯匙
鹽半茶匙
糖 1/4 茶匙
豬油 1 湯匙
水 1 1/2 杯
300 g spinach
300 g wheat starch
1 tbsp Thai tapioca starch
2 tbsp caltrop starch
1/2 tsp salt
1/4 tsp sugar
1 tbsp lard
1 1/2 cup water

餡料 Filling

鮮蝦肉 8 兩
冬筍 10 兩
肥豬肉 1 湯匙
300 g shelled shrimps
375 g bamboo shoot
1 tbsp fatty pork

調味料 Filling seasoning

鹽 3/4 茶匙
糖半茶匙
麻油半茶匙
生粉 1 湯匙
胡椒粉少許
3/4 tsp salt
1/2 tsp sugar
1/2 tsp sesame oil
1 tbsp caltrop starch
ground white pepper

Har Gow with spinach skin

外皮做法 Method for dumpling skin dough

1. 澄麵、生粉、薯粉、鹽及糖篩勻，備用。

2. 菠菜洗淨，切碎，放入攪拌機內，加水 1 1/2 杯打成茸，隔渣，餘下菠菜汁 2 杯（若份量不足，加水補充）。

3. 煮滾菠菜汁，下豬油略煮，加入粉料，用木棒迅速拌勻，熄火，加蓋焗 5 分鐘。

1. Sieve wheat starch, caltrop starch, tapioca starch, sugar and salt.

2. Rinse spinach and finely chop it. Transfer into a blender. Add 1 1/2 cups of water and blend until fine. Strain to yield 2 cups of spinach juice. (Thin it out to 2 cups by adding water if there's not enough).

4. Boil spinach juice and add lard to cook briefly. Add dry ingredients and mix well with a wooden spatula. Turn off the heat. Cover the lid and leave it for 5 minutes.

餡料做法 Method for filling

1. 鮮蝦肉去腸，加生粉及鹽各半茶匙醃 5 分鐘，洗淨，抹乾。

2. 冬筍洗淨，飛水，切絲，壓乾水分；肥豬肉烚熟，切粒。

3. 將餡料與調味料拌勻，攪至起膠，放入雪櫃待片刻。

1. Devein the shrimps. Add 1/2 tsp of caltrop starch and 1/2 tsp of salt. Mix well and leave them for 5 minutes. Rinse and wipe dry.

2. Rinse bamboo shoot and blanch in boiling water. Shred finely and squeeze dry. Boil the fatty pork in water until done. Dice finely.

3. Put all ingredients into a mixing bowl. Add filling seasoning and stir until sticky. Refrigerate for a while before use.

包餡做法 Assembly

1. 桌面灑上薯粉，將麵糰搓成軟滑粉糰，切成小粒，用擀麵棒擀薄，壓成窩形餃子皮。

2. 將後方的皮打摺，捏成窩形狀，包入適量餡料，再將另一邊的皮覆上，推捏成蝦餃形狀。

3. 排入已塗油蒸籠內，隔水以大火蒸 7 分鐘即可。

1. Dust counter with tapioca starch. Knead into smooth dough. Divide into small pieces. Roll each piece out into a thin round disc with its rim curling up.

2. Pleat the rim of the dumpling skin closer to you so that it looks like a bowl. Put in some filling. Fold the far end of the dumpling skin toward you. Pinch and press firmly into a typical crescent shape.

3. Arrange the dumplings in a greased steamer. Steam for 7 minutes. Serve.

上湯菜肉餃 Bok choy and pork dumplings in stock

材料 Ingredients

白菜 1 1/2 斤
梅頭豬肉 12 兩
大頭菜 2 片
葱粒少許
清雞湯 1 罐（250ml）
白色方形水餃皮 1 斤
900 g bok choy
450 g pork shoulder butt
2 pieces salted kohlrabi
diced spring onion
1 can (250 ml) chicken stock
600 g round white dumpling skin

調味料 Seasoning

鹽 2 茶匙
糖半茶匙
生抽 1 茶匙
生粉 2 湯匙
蛋白半個
麻油及胡椒粉各少許
2 tsp salt
1/2 tsp sugar
1 tsp light soy sauce
2 tbsp caltrop starch
1/2 egg white
sesame oil
ground white pepper

上湯 Stock

清雞湯 6 杯
老抽 2 湯匙
糖半茶匙
鹽適量
麻油 2 湯匙
6 cup chicken stock
2 tbsp dark soy sauce
1/2 tsp sugar
salt
2 tbsp sesame oil

做法 Method

1. 白菜洗淨，飛水，用水沖涼，切碎，壓乾水分。
2. 豬肉洗淨，剁茸；大頭菜洗淨，切粒。
3. 白菜、豬肉、大頭菜及葱粒拌勻，下調味料攪拌均勻，備用。
4. 餃皮包入適量餡料，對摺，用水封口，揑成半圓形，兩端重疊並揑緊，呈帽子狀。
5. 燒滾水，放入餃子煮至浮起，加入清水半碗再煮至餃子浮面，盛起。
6. 煮滾上湯，放入餃子，灑上葱粒即成。

1. Rinse the bok choy. Blanch in boiling water. Rinse in running tap water until cold. Finely chop it. Squeeze dry and set aside.

2. Rinse the pork and finely chop it. Rinse the salted kohlrabi and dice it.

3. Mix bok choy, pork, salted kohlrabi and spring onion together. Add seasoning and mix again.

4. Put some filling on a piece of dumpling skin. Wet the rim with water. Fold in half and press firmly. Then fold the two pointy ends together.

5. Boil water in a pot. Cook dumplings in boiling water until they float. Add 1/2 bowl of cold water. Cook until the dumplings float again. Drain and set aside.

6. Boil the stock. Put in the dumplings. Sprinkle with diced spring onion. Serve.

小煮意

當水餃煮至浮面時，加半碗清水再煮滾，令水餃爽口之餘，餡料容易熟透。

When the dumplings float on the boiling water the first time, it may not be cooked through. By adding 1/2 bowl of cold water and letting it boil again, you can make the dumplings more crunchy in texture and the filling gets cooked more easily.

香滑蝦米腸粉

Rice noodle roll with dried shrimps

材料 Ingredients

粘米粉 16 安士（量杯計）
泰國薯粉 3 湯匙
粟粉 2 湯匙
馬蹄粉 1 湯匙
油 2 湯匙
鹽 1 茶匙
蝦米粒 2 兩
葱 3 棵（切粒）
清水 4 杯

470 ml long-grain rice flour
(measured in cup)
3 tbsp Thai tapioca starch
2 tbsp cornstarch
1 tbsp water chestnut starch
2 tbsp oil
1 tsp salt
75 g diced dried shrimps
3 sprigs spring onion (diced)
4 cups water

蘸醬 Dipping sauce

芝麻醬 6 湯匙
海鮮醬（甜醬）4 湯匙
熟油 6 湯匙
芝麻少許
* 拌勻
6 tbsp sesame paste
4 tbsp Hoi Sin sauce
6 tbsp cooked oil
sesames
* mixed well

調味汁 Seasoning

老抽及生抽各 2 湯匙
糖 1 茶匙
胡椒粉少許
滾水 2 湯匙
* 拌勻
2 tbsp light soy sauce
2 tbsp dark soy sauce
1 tsp sugar
ground white pepper
2 tbsp boiling water
* mixed well

❧ 小煮意 ❧

- 白洋布於布匹店有售，沾濕後使用，每次蒸完後，必須搓洗乾淨才再使用。

- 每次舀粉漿前，必須先將粉漿攪勻，將一勺粉漿倒入盆內，別補加粉漿，否則腸粉表面不美觀。

- 若倒入的粉漿不平，宜搖動蒸盆令粉漿均勻。

- You can get white muslin from any fabric store. Make sure you wet it before use. Rub it and rinse it really well after use.

- Before you ladle the rice batter onto the muslin cloth, make sure you stir the batter well. You should always pour enough batter on the cloth in one go. Adding batter later makes the noodles look patchy and not very pleasing.

- If the rice batter doesn't flow to smooth itself out, jiggle and swirl the tray a little to coat evenly.

做法 Method

1. 粘米粉、薯粉、粟粉、馬蹄粉、鹽、油及清水徹底拌勻,用密篩過濾兩次成幼滑米漿。

2. 取白洋布沾濕,鋪在不銹鋼盆上,倒入適量米漿,趁米漿未熟時,快手放入蝦米及葱粒,加蓋,隔水蒸約 3 分鐘。

3. 取出白洋布,反轉覆在已鋪保鮮紙或塗油的桌面,輕輕拉起白洋布,再捲成腸粉,伴蘸醬及調味汁享用。

1. Mix rice flour, tapioca starch, cornstarch, water chestnut starch, salt, oil and water until well incorporated. Pass the rice batter through a fine sieve twice.

2. Wet a sheet of muslin cloth. Line a stainless steel tray with the damp cloth. Pour some rice batter over the muslin. Put on some dried shrimps and spring onion before it sets. Cover the lid and steam for 3 minutes.

3. Take the muslin cloth out of the steamer. Flip it upside down onto a counter either lined with cling film or greased with oil. Gently peel off the muslin cloth. Roll the noodles into cylinders. Serve with dipping sauce and seasoning.

Imitation shark's fin soup

材料 Ingredients

瘦肉 3 兩
罐裝冬筍 1 罐
木耳半兩
冬菇 6 朵
粉絲 1 扎（細）
上湯 1 盒（約 500ml）
清水 5 碗
113 g lean pork
1 can bamboo shoots
19 g wood ear fungus
6 dried shiitake mushrooms
1 small bundle mung bean vermicelli
1 pack stock (about 500ml)
5 bowls water

蒸冬菇調味料
Seasoning for shiitake mushrooms

生抽、糖、油及紹酒各 1 茶匙
浸冬菇水 2 湯匙
1 tsp light soy sauce
1 tsp sugar
1 tsp oil
1 tsp Shaoxing wine
2 tbsp water in which
shiitake mushrooms are soaked

獻汁 Thickening glaze

馬蹄粉 4 湯匙
生抽 2 茶匙
老抽 1 湯匙
糖、麻油、胡椒粉、紹酒及鹽各少許
4 tbsp water chestnut starch
2 tsp light soy sauce
1 tbsp dark soy sauce
sugar
sesame oil
ground white pepper
Shaoxing wine
salt

❧ 小煮意 ❧

• 瘦肉可先焓熟，或取煲湯後的瘦肉，待涼後用鹽醃 3 天使用。

• 用馬蹄粉埋獻，目的是令湯汁濃郁，而且透明好看，時間久了也不會稀稀的。

• Alternatively, you may blanch the pork in water until done, or use lean pork after you've made soup with it. Let cool and rub salt on it. Leave it for 3 days.

• Thick soup thickened with water chestnut starch looks transparent and stays thick for a long time, whereas soup thickened with other starches tend to turn watery after a while.

做法 Method

1. 瘦肉用鹽 3 湯匙醃 3 天,取出,放入滾水內煲約 15 分鐘,用手撕成肉絲。

2. 冬菇浸軟,去蒂,加入冬菇調味料蒸 15 分鐘,切絲。

3. 冬筍飛水,切幼絲,擠乾水分;粉絲浸軟,切段;木耳浸軟後切幼絲。

4. 燒熱鑊,下少許油,灒酒,加入上湯及清水略煮,下所有材料煮約 10 分鐘,埋獻,略拌即成,享用時可加入麻油、胡椒粉或浙醋。

1. Rub 3 tbsp of salt on the pork. Leave it to marinate for 3 days. Boil the pork in water for about 15 minutes. Tear the pork into shreds with your hands.

2. Soak shiitake mushrooms in water until soft. Cut off the stems. Add seasoning for shiitake mushrooms and steam for 15 minutes. Finely shred.

3. Blanch bamboo shoots in boiling water. Drain and finely shred them. Squeeze dry. Soak the mung bean vermicelli in water until soft. Cut into short lengths. Soak wood ear fungus in water until soft. Finely shred it.

4. Heat the wok and add a little oil. Sizzle with wine. Add stock and water. Bring to the boil. Put in all ingredients and cook for 10 minutes. Stir in the thickening glaze. Mix well and serve with sesame oil, ground white pepper and red vinegar on the side.

越南春卷 Vietnamese spring rolls

材料 Ingredients

春卷皮 4 安士
瘦肉 4 安士
蝦肉 4 安士
甘筍 1 個（小）
木耳少許
蝦米半兩
粉絲半兩
112 g spring roll skin
112 g lean pork
112 g shelled shrimps
1 small carrot
wood ear fungus
19 g dried shrimps
19 g mung bean vermicelli

醃料 Marinade

鹽 3/4 茶匙
糖、魚露及生粉各半茶匙
3/4 tsp salt
1/2 tsp sugar
1/2 tsp fish sauce
1/2 tsp caltrop starch

調味料 Seasoning

生抽及魚露各 1 茶匙
鹽半茶匙
糖 1 茶匙
生粉半茶匙
水 2 安士
1 tsp light soy sauce
1 tsp fish sauce
1/2 tsp salt
1 tsp sugar
1/2 tsp caltrop starch
60 ml water

蘸汁 Dipping sauce

紅椒 1 隻（切粒）
芫茜 1 棵（切碎）
薑米及蒜米各 2 湯匙
白醋及魚露各 2 安士
糖 6 茶匙
* 拌勻
1 red chilli (diced)
1 sprig coriander (finely chopped)
2 tbsp diced ginger
2 tbsp diced garlic
60 ml white vinegar
60 ml fish sauce
6 tsp sugar
* mixed well

∾ 小煮意 ∾

- 春卷皮冷藏後較硬，用啤酒或糖水略浸，可軟化餅皮。

- 我特別喜歡用啤酒浸春卷皮，帶酒香味。

- Spring roll skin tends to be a bit stiff after refrigerated. You may soften it up by soaking them in beer or syrup briefly.

- I especially prefer soaking spring roll skin in beer. The spring roll has a mild alcoholic aroma after fried.

做法 Method

1. 瘦肉及甘筍切絲；木耳浸軟、切絲；蝦米切粒；粉絲浸軟、切碎。
2. 瘦肉及蝦肉用醃料拌勻，待一會。
3. 燒熱少許油，加入瘦肉及蝦肉炒熟，下其餘材料炒勻，拌入調味料，盛起，待涼。
4. 春卷皮用啤酒略浸至軟身，包入一份材料成長條形，下熱油內炸至金黃色，伴蘸汁享用。

1. Shred the pork and carrot. Soak the wood ear in water until soft. Finely shred it. Dice the dried shrimps. Soak the mung bean vermicelli in water until soft. Cut into short lengths.

2. Add marinade to the pork and shrimps. Mix well and leave them for a while.

3. Heat oil in a wok. Stir fry pork and shrimps until done. Put in remaining ingredients. Toss well. Add seasoning and set aside to let cool. This is the filling.

4. Soak the spring roll skin in beer until soft. Put some filling over it. Roll the spring roll skin up into a cylindrical parcel. Deep fry in hot oil until golden. Serve with the dipping sauce on the side.

Dace balls and glutinous rice balls in stock 鹹湯丸

材料 Ingredients

冬菇 8 朵
蘿蔔半斤
黃芽白半斤
芹菜粒 6 湯匙
葱粒少許
8 dried shiitake mushrooms
300 g white radish
300 g Napa cabbage
6 tbsp diced Chinese celery
diced spring onion

鯪魚肉材料
Ingredients for dace balls

鯪魚肉 4 兩
蝦米粒 3 湯匙
臘肉茸 3 湯匙
葱粒 1 湯匙
鹽半茶匙
胡椒粉少許
生粉 1 湯匙
水 1 湯匙
150 g minced dace
3 tbsp diced dried shrimps
3 tbsp diced Chinese preserved pork belly
1 tbsp diced spring onion
1/2 tsp salt
ground white pepper
1 tbsp caltrop starch
1 tbsp water

湯丸材料
Ingredients for glutinous rice balls

糯米粉 3 杯
鹽半茶匙
水 11 安士
3 cups glutinous rice flour
1/2 tsp salt
325 ml water

上湯調味料 Stock seasoning

上湯 8 杯
金華火腿茸 3 湯匙
鹽適量
糖、麻油、胡椒粉及紹酒各少許
8 cups stock
3 tbsp grated Jinhua ham
salt
sugar
sesame oil
ground white pepper
Shaoxing wine

小煮意

金華火腿鹹味重，可先飛水 15 分鐘，沖洗
後加入糖及紹酒蒸 15 分鐘，去掉鹹味及油
膩味道。

Jinhua ham is quite salty. You may blanch it in
boiling water for 15 minutes first. Then rinse
in water and steam it with some sugar and
Shaoxing wine for 15 minutes before use. That
would make the ham less salty and greasy.

做法 Method

1. 將鯪魚肉材料拌勻，攪打至起膠，搓成魚丸狀，備用。
2. 冬菇用水浸軟，去蒂，切絲；蘿蔔及黃芽白洗淨，切絲。
3. 燒熱少許油，下冬菇、蘿蔔及黃芽白炒香，放入上湯及金華火腿茸煮片刻，下魚丸及上湯調味料煮熟，盛起材料，上湯留用。
4. 糯米粉篩勻，放於大碗內，中間開一穴，將鹽及水分數次加入，搓成軟滑的粉糰，再搓成小粒狀。
5. 小湯丸逐粒放入上湯內，煮至浮起，再放入步驟3的材料略煮，最後下芹菜粒及葱粒滾一會即可。

1. Mix all dace ball ingredients together. Stir until sticky. Lift it and slap it a few times into a bowl. Roll into dace balls. Set aside.

2. Soak shiitake mushrooms in water until soft. Cut off the stems. Shred them. Rinse radish and cabbage. Shred them.

3. Heat some oil in a wok. Stir fry mushrooms, radish and cabbage until fragrant. Add stock and grated Jinhua ham. Cook briefly. Put in dace balls and stock seasoning. Cook until dace balls are done. Set aside all solid ingredients. Save the stock for later use.

4. To make glutinous rice balls, sieve the glutinous rice flour into a mixing bowl. Make a well at the centre. Put in some salt and water each time and mix well after each addition. Knead into smooth dough. Then roll into small spheres.

5. Bring the stock from step 3 to the boil. Put in the glutinous rice balls and cook till they float. Put the solid ingredients from step 3 back in. Add diced Chinese celery and spring onion. Boil briefly. Serve.

自製蜜汁豬肉乾

Home-made honey-glazed pork jerky

～ 小煮意 ～

依此烘烤方法，也可焗製牛肉乾或雞肉乾。
You may also make beef or chicken jerky with the same recipe.

材料 Ingredients

免治豬肉 12 兩
蜜糖 2 湯匙
檸檬汁 1 湯匙
450 g ground pork
2 tbsp honey
1 tbsp lemon juice

調味料 Seasoning

玫瑰露酒半湯匙
油 1 湯匙
生抽 1 茶匙
老抽 1 湯匙
糖 5 茶匙
魚露 1 1/2 湯匙
胡椒粉少許
1/2 tbsp Chinese rose wine
1 tbsp oil
1 tsp light soy sauce
1 tbsp dark soy sauce
5 tsp sugar
1 1/2 tbsp fish sauce
ground white pepper

做法 Method

1. 免治肉與調味料拌勻，攪至起膠，放入雪櫃冷藏 1 小時。

2. 桌面鋪上保鮮紙，放上半份免治肉抹平，蓋上另一層保鮮紙，用擀麵棒壓平至 1/2cm 厚，轉放已鋪牛油紙的焗盆內。

3. 焗爐調至 200℃，預熱 5 分鐘，放入免治肉烘 12 分鐘，取出，塗抹蜜糖及檸檬汁再焗 3 分鐘，取出，再掃蜜糖及檸檬汁焗 2 分鐘，將肉片反轉，依上述方法塗抹及烘焗即成。

1. Mix ground pork with seasoning. Stir until sticky. Refrigerate for 1 hour.

2. Spread a sheet of cling film on the counter. Put on 1/2 of the ground pork and smooth it out. Put another sheet of cling film over it. Roll the pork out until 1/2 cm thick. Remove the cling film and transfer onto a baking tray lined with baking paper. Repeat this step with the remaining pork.

3. Preheat an oven to 200°C for 5 minutes. Bake the pork in the oven for 12 minutes. Brush honey and lemon juice over it and bake for 3 more minutes. Brush honey and lemon juice on it again. Bake for 2 minutes. Flip the pork and repeat this step again. Serve.

白雲豬手

Cold pork trotters in pickle brine

材料 Ingredients

豬手 2 隻（斬件）
長紅椒 2 隻（切片）
蒜茸 1 茶匙
2 pork trotters (chopped into pieces)
2 red chillies (sliced)
1 tsp grated garlic

調味料 Seasoning

白米醋 1 瓶（大）
砂糖 12 兩
鹽 2 茶匙
1 large bottle rice vinegar
450 g sugar
2 tsp salt

❧ 小煮意 ❧

煮滾調味料後，必須待冷才放入豬手，浸泡後的肉質才爽口。

After you boil the seasoning, make sure you let it cool completely before putting the pork trotters in. This way, the meat would retain its lovely chew after being marinated.

做法 Method

1. 豬手放入滾水內，加入白醋 1 湯匙，飛水，待約 10 分鐘。
2. 燒滾一鍋水，放入豬手煲至軟腍，取出，用冰水沖洗，洗掉油分，再浸泡冰水 1 小時，瀝乾水分。
3. 將調味料放入瓦煲內煮滾，待冷，放入豬手、紅椒及蒜茸，浸約 6 至 8 小時即可。

1. Blanch the pork trotters in a pot of boiling water. Add 1 tbsp of white vinegar. Cook for about 10 minutes. Drain.

2. Boil another pot of water and cook the pork trotter until tender. Drain and rinse in ice water to remove the grease. Soak in ice water for 1 hour. Drain well.

3. Put all seasoning into a clay pot. Bring to the boil. Let cool. Put in the cooked pork trotters, red chillies and grated garlic. Soak for 6 to 8 hours. Serve.

葱油餅　Spring onion pancake

皮料 Ingredients for pastry

麵粉 8 安士（量杯計）
暖水 6 至 8 湯匙
豬油 2 湯匙
鹽 3/4 茶匙
糖 1/4 茶匙

235 ml flour (measured in cup)
6 to 8 tbsp warm water
2 tbsp lard
3/4 tsp salt
1/4 tsp sugar

餡料 Filling

葱 4 棵
鹽半茶匙

4 sprigs spring onion
1/2 tsp salt

∾ 小煮意 ∾

- 建議用豬油搓成粉糰，別用其他油代替，否則煎後的葱油餅香氣不足。
- 於包餡料前，才將葱粒與鹽拌勻，否則太早拌在一起，容易溢出水分。
- 別太用力搓麵糰，否則麵糰起筋，進食時口感太硬。

- I prefer using lard in this recipe. Other oils just won't taste as aromatic.
- Mix the spring onion with salt only right before you wrap them in the dough. Mixing them too early will draw the water out of the spring onion and the filling will be too watery.
- Do not knead the dough too hard. Otherwise, the gluten may over-develop and the pancake will be too hard after cooked.

做法 Method

1. 葱洗淨，切去頭尾，切幼粒，下鹽拌勻，備用。

2. 麵粉篩勻於桌面，中間開一穴，放入糖及鹽和勻麵粉，下豬油及暖水拌勻，搓成軟粉糰，用濕布蓋着待片刻。

3. 將粉糰分成四等份，用擀麵棒擀成 4 吋 x 6 吋方塊，葱粒排於中央，兩邊向內摺起，再捲成蛇餅狀，用水收口，略壓平。

4. 燒紅鑊，下油半碗，放入麵糰用慢火煎至兩邊金黃色，趁熱享用。

1. Rinse the spring onion. Cut off the roots and the tips. Dice finely. Add salt and stir well.

2. Sieve the flour on a counter. Make a well at the centre. Put in sugar and salt. Mix well. Add lard and warm water. Knead into smooth dough. Cover with damp cloth for a while.

3. Divide the dough into 4 equal portions. Roll each out into a 4-by-6 inch rectangle with a rolling pin. Put the spring onion filling at the centre. Roll both sides toward the centre. Then coil it. Wet the seam with water. Press gently to flatten it.

4. Heat a wok and add 1/2 bowl of oil. Fry the dough over low heat until both sides golden. Serve hot.

材料 Ingredients

芋頭 10 兩
腐皮 3 張
澄麵 3 湯匙

375 g taro
3 sheets beancurd skin
3 tbsp wheat starch

調味料 Seasoning

鹽 1 1/2 茶匙
糖半茶匙
油 2 湯匙
胡椒粉及素味精各少許

1 1/2 tsp salt
1/2 tsp sugar
2 tbsp oil
ground white pepper
vegetarian MSG

香荔酥卷

Beancurd skin roll with taro filling

做法 Method

1. 芋頭去皮、切件，蒸 30 分鐘至熟，搓爛成芋茸，下少許糖、鹽及油 2 湯匙拌勻。

2. 煮滾水 4 湯匙，下澄麵攪勻，加蓋焗片刻，取出，與芋茸搓勻，備用。

3. 腐皮一張鋪於桌面，放上芋茸塗勻，再蓋上另一張腐皮，捲成春卷狀，再包上第三張腐皮，包捲，接口用生粉漿黏緊。

4. 芋卷切成塊狀，放入滾油內炸至金黃色，切件上碟，蘸甜酸醬享用。

1. Peel and cut taro into pieces. Steam for 30 minutes until done. Mash the taro. Add a pinch of sugar, salt and 2 tbsp of oil. Mix well.

2. Boil 4 tbsp of water. Add wheat starch and mix well. Cover the lid and leave it for a while. Take out. Mix with mashed taro.

3. Spread a sheet of beancurd skin on the counter. Spread some mashed taro filling over it. Cover with another sheet of beancurd skin. Roll it up into a cylinder. Then wrap it the third sheet of beancurd skin. Seal the seam with some caltrop starch slurry.

4. Cut the beancurd skin roll into pieces. Deep fry in hot oil until golden. Save on a serving plate. Serve with sweet and sour sauce as a dip.

荔茸葱油餅

Taro spring onion pancakes

皮料 Ingredients for pastry

芋頭 12 兩	450 g taro
麵粉 2 杯	2 cups flour
芝麻 1 湯匙	1 tbsp sesames
葱 3 兩	113 g spring onion
鹽 2 茶匙	2 tsp salt
糖 3/4 茶匙	3/4 tsp sugar
暖水 10 至 12 湯匙	10 to 12 tbsp warm water
豬油 3 湯匙	3 tbsp lard

❧ 小煮意 ❧

購買芋頭時，選較輕身及芋頭肉呈白色的，口感較粉糯，蒸熟後容易壓成芋茸。

When you shop for taro, pick those that feels light in your hand and those that looks white and powdery on the cut. Such taros are more starchy and taste better after mashed.

做法 Method

1. 葱洗淨，去頭尾，切成幼粒，下鹽半茶匙拌勻。
2. 芋頭洗淨、切片，隔水蒸 15 分鐘，取出，壓成芋茸。
3. 麵粉篩勻桌面上，中間開一穴，放入芋茸、糖、鹽及豬油，慢慢加入暖水搓勻，下葱粒搓成軟滑粉糰，用布蓋上待 15 分鐘。
4. 將麵糰分成兩份，擀薄成圓形薄片，灑上少許芝麻。
5. 燒熱平底鑊，下半碗油用慢火煎至兩面微黃色，加入半碗水，加蓋，以慢火煎至水乾透及葱油餅呈金黃色，上碟享用。

1. Rinse the spring onion. Cut off the roots and the tips. Dice finely. Add 1/2 tsp of salt and stir well.

2. Rinse and slice the taro. Steam for 15 minutes. Mash it with a spoon.

3. Sieve the flour on a counter. Make a well at the centre. Put in mashed taro, sugar, salt and lard. Slowly add warm water and mix well. Add spring onion and knead into smooth dough. Cover with damp cloth for 15 minutes.

4. Divide the dough into 2 equal portions. Roll each out into a round disc with a rolling pin. Sprinkle with some sesames on top.

5. Heat a wok and add 1/2 bowl of oil. Fry the pancake over low heat until both sides golden. Add 1/2 bowl of water. Cover the lid and cook over low heat until water dries out and the pancake turns golden. Serve.

蝦肉木耳腐皮夾

Beancurd skin sandwich with minced shrimp and wood ear

材料 Ingredients

腐皮 2 張
蝦肉 4 兩
鯪魚肉 4 兩
木耳半塊
蔥白 2 湯匙
2 sheets beancurd skin
150 g shelled shrimps
150 g minced dace
1/2 piece wood ear fungus
2 tbsp white part of spring onion (diced)

調味料 Seasoning

鹽 3/4 茶匙
生粉 1 茶匙
蛋白 2 湯匙
胡椒粉少許
3/4 tsp salt
1 tsp caltrop starch
2 tbsp egg white
ground white pepper

做法 Method

1. 木耳用水浸透及脹發，剪去硬蒂，切成幼絲。

2. 蝦肉去腸，下生粉及鹽各 1 茶匙拌勻待 5 分鐘，洗淨，用布包裹壓乾水分，用刀面壓成蝦茸，略剁，加入鯪魚肉、木耳及調味料拌勻，攪打至起膠。

3. 腐皮用濕布抹淨，切成四方型（約 3 吋 x 4 吋），放入餡料 2 湯匙抹平，左右兩邊覆入，用生粉水黏口。

4. 燒熱鑊下油，放入腐皮夾用慢火煎至兩面金黃色，瀝去油分，上碟，蘸喼汁享用。

1. Soak the wood ear in water until soft. Cut off the root. Cut into finely shreds.

2. Devein the shrimps. Add 1 tsp of salt and 1 tsp of caltrop starch. Mix well. Leave them for 5 minutes. Rinse and wrap them in dry towel. Squeeze and wipe dry. Transfer onto a chopping board. Crush the shrimps with the flat side of a knife. Finely chop it. Add minced dace, wood ear and seasoning. Mix well. Stir until sticky. Lift the mixture up and slap it on the chopping board a few times. This is the filling.

3. Wipe the beancurd skin with damp towel. Cut into rectangle about 3 by 4 inches. Put 2 tbsp of filling on it. Smear flat. Fold both sides of the beancurd skin toward the centre. Seal the seam with caltrop starch slurry.

4. Heat oil in a wok. Fry the beancurd skin sandwiches over low heat until both sides golden. Drain and save on a serving plate. Serve with the Worcestershire sauce.

小煮意

腐皮夾可煎吃或放入上湯內煮熱，具不同的食味感受。

The beancurd skin sandwiches can be pan-fried or poached in stock. They taste equally great either way.

醃雲耳蘿蔔皮

Radish skin and cloud ear fungus pickles

材料 Ingredients

蘿蔔 2 個
雲耳 2 錢
白米醋半斤
砂糖約 7 兩
鹽 1 茶匙
2 white radishes
8 g cloud ear fungus
300 g rice vinegar
260 g sugar
1 tsp salt

做法 Method

1. 蘿蔔洗淨，抹乾水分，刨皮，將蘿蔔皮放於大碗內，加入鹽 1 茶匙醃約 1 小時，擠乾水分。

2. 白米醋煮熱，下砂糖煮至糖溶，待涼備用。

3. 雲耳浸軟，洗淨，灼熟，吸乾水分。

4. 蘿蔔皮用凍開水洗去鹽分，吸乾水分，與雲耳一併放入盛有米醋糖水的玻璃瓶內醃約 2 小時即可。

1. Rinse the radishes and wipe dry. Peel them and save the peel in a large mixing bowl. Add 1 tsp of salt and mix well. Leave it for 1 hour. Squeeze the water.

2. Boil the vinegar in a clay pot. Add sugar and cook until sugar dissolves. Leave it to cool.

3. Soak cloud ear fungus in water until soft. Rinse and blanch in boiling water until done. Drain.

4. Rinse the radish peel with cold drinking water to remove the salt. Wipe dry. Put radish peel and cloud ear fungus into the vinegar syrup in a glass bottle. Leave them for 2 hours. Serve.

❧ 小煮意 ❧

必須待米醋糖水涼透，才放入雲耳及蘿蔔皮，否則米醋的熱度令材料的質感不爽脆。

Make sure the vinegar syrup is completely cooled before putting in the cloud ear and radish peel. Otherwise, the residual heat in the syrup may cook the ingredients, so that the pickles won't be crunchy.

Chaozhou pan-fried dried radish cakes

潮州煎菜頭粿

皮料 Ingredients for dough

糯米粉 16 安士（量杯計）
澄麵 4 湯匙
鹽半茶匙
豬油 1 湯匙
白芝麻適量
清水 6 安士
滾水 2 安士

470 ml glutinous rice flour
(measured in cup)
4 tbsp wheat starch
1/2 tsp salt
1 tbsp lard
white sesames
175 ml water
60 ml boiling water

調味料 Seasoning

鹽及糖各半茶匙
生抽 2 茶匙
雞 1/4 茶匙
麻油及胡椒粉各少許

1/2 tsp salt
1/2 tsp sugar
2 tsp light soy sauce
1/4 tsp chicken bouillon powder
sesame oil
ground white pepper

餡料 Filling

梅頭瘦肉 4 兩
蝦肉 4 兩
芹菜 3 棵
馬蹄 6 粒（去皮）
菜頭（蘿蔔乾）1 兩
蝦米 2 湯匙

150 g pork shoulder butt
150 g shelled shrimps
3 sprigs Chinese celery
6 water chestnuts (peeled)
38 g dried radish
2 tbsp dried shrimps

生粉獻 Caltrop starch slurry

生粉 1 茶匙
水 1 湯匙
* 拌勻

1 tsp caltrop starch
1 tbsp water
* mixed well

❧ 小煮意 ❧

• 潮汕地區人士稱蘿蔔乾為菜頭，雜貨店
有售。

• 糯米粉與熟澄麵搓勻，令麵糰煙韌彈軟。

• 搓成的菜頭粿，無論煎、炸或蒸，皆美
味可口。

• You can get dried radish from traditional
grocery stores.

• Mixing glutinous rice flour and wheat
starch in the dough helps keep the cake
chewy without being overly sticky.

• Apart from pan-frying, the dried radish
cakes also taste after deep-fried or steamed.

外皮做法 Method for dough

1. 糯米粉篩勻，放入大碗內，分數次加入清水，搓成軟麵糰。
2. 澄麵篩勻，放於另一大碗內，拌入滾水拌勻，加蓋焗片刻。
3. 澄麵糰與糯米粉糰搓勻，加入豬油及鹽搓至幼滑，用濕毛巾蓋着。

1. Sieve the glutinous rice flour into a mixing bowl. Add a little water at a time while mixing continuously. Knead into soft dough.

2. Sieve wheat starch into another bowl. Stir in boiling water and mix well. Cover the lid and leave it for a while.

3. Knead the wheat starch dough with the glutinous rice dough. Add lard and salt. Knead till smooth. Cover with damp towel.

做法 Assembly

1. 所有餡料洗淨，切幼粒。燒熱鑊，放入餡料炒勻，下調味料拌炒，埋獻，上碟待涼。
2. 軟麵糰搓成長條，分成小粒，用手搓成窩形，包入餡料，收口後略按壓，灑上白芝麻。
3. 燒熱平底鑊，下少許油，放入小圓餅煎至兩面金黃色即可。

1. Rinse all filling ingredients. Dice finely. Heat a wok and stir fry filling until well mixed. Add seasoning and stir again. Stir in caltrop starch slurry. Set aside to let cool.

2. Roll the dough into a long cylinder. Cut into small pieces. Roll each piece of dough with your hands into a little bowl. Put in some filling and seal the seam. Press gently to flatten into a round patty. Coat the cake in white sesames.

3. Heat a pan and add some oil. Fry the round patties until both sides golden. Serve.

香醬汁雞翼

Deep-fried chicken wings in nutty fruity soy sauce

材料 Ingredients

雞中翼 1 磅
菠蘿 3 圓片（剁幼）
450 g chicken mid-joint wings
3 rings canned pineapple
(finely chopped)

醃料 Marinade

雞蛋 1 個
鹽及糖各半茶匙
生抽 1 湯匙
雞粉半茶匙
1 egg
1/2 tsp salt
1/2 tsp sugar
1 tbsp light soy sauce
1/2 tsp chicken bouillon powder

醬汁料 Sauce

老抽 1 茶匙
鹽 1/4 茶匙
花生醬 2 湯匙
糖 2 茶匙
清水半碗
1 tsp dark soy sauce
1/4 tsp salt
2 tbsp peanut butter
2 tsp sugar
1/2 bowl water

做法 Method

1. 雞翼解凍，洗淨，抹乾水分，拌入醃料待 30 分鐘。
2. 燒熱油，雞翼撲上粟粉，放入滾油內炸至金黃色，隔去油分，上碟。
3. 燒熱油 1 湯匙，下醬汁料煮滾，最後下菠蘿茸，蘸雞翼享用。

1. Thaw the chicken wings and rinse well. Wipe dry. Stir in marinade and leave them for 30 minutes.
2. Heat oil in a wok. Coat chicken wings lightly in cornstarch. Deep fry until golden. Drain and arrange on a serving plate.
3. Heat 1 tbsp of oil in a wok. Pour in all sauce ingredients and bring to the boil. Stir in chopped pineapple at last. Serve this sauce as a dip on the side with the chicken wings.

小煮意

- 鍾情脆口的雞翼嗎？雞翼調味後，放在陰涼處吹乾，最後下油鍋炸脆即可。

- 冰鮮雞翼加入雞粉醃一會，可辟除騷味及冷藏味。

- For ultra-crispy chicken wings, leave them in the shade to air-dry after you marinate them. The wings will turn out crispier after fried.

- Adding chicken bouillon powder to marinate frozen chicken wings helps remove the gamey taste and freezer odour.

魷魚鬚雜菜天婦羅

Tempura squid tentacles and assorted veggies

小煮意

用冰水調成的炸粉漿，比用清水開調的更鬆脆可口。一般超市有售。

Using ice water in the deep-frying batter makes the tempura more crispy and light than using tap water. It is available at the supermarket.

材料 Ingredients

魷魚鬚 3 兩
椰菜絲半碗
青、紅甜椒絲半碗
西芹絲 1/4 碗
粟米粒 1/4 碗
甘筍絲 1/4 碗
炸粉 1 杯
冰水適量
113 g squid tentacles
1/2 bowl shredded white cabbage
1/2 bowl shredded green and red bell pepper
1/4 bowl shredded celery
1/4 bowl sweet corn kernels
1/4 bowl shredded carrot
1 cup deep-frying batter mix
ice water

做法 Method

1. 魷魚鬚洗淨，放入滾水內略灼後即盛起，抹乾水分，切成細段。
2. 所有雜菜絲洗淨，瀝乾，拌入魷魚鬚，加入胡椒粉、少許鹽及糖備用。
3. 炸粉放入碗內，加入少許鹽及冰水調成糊狀，將油 1 湯匙放糊漿面，冷藏約 15 分鐘，拌成脆漿。
4. 將所有材料放入脆漿內拌勻，以湯杓盛起，放入滾油內炸至金黃色（連湯杓一起炸），至天婦羅脫離即成。

1. Rinse the squid and blanch in boiling water briefly. Drain and wipe dry. Cut into short lengths.

2. Rinse all shredded veggies and wipe dry. Put in squid and mix well. Season with salt, ground white pepper and sugar.

3. Put the deep-frying batter mix into a bowl. Add a pinch of salt and ice water. Stir into a runny paste. Pour 1 tbsp of cooking oil over the paste. Refrigerate for 15 minutes. Mix well.

4. Put all solid ingredients into the batter. Stir well. Transfer a ladle of the mixture into hot oil and deep fry until golden. Keep the ladle in the oil until the tempura is set and separates from the ladle. Drain and serve.

臘味蘿蔔糕 Radish cake with preserved sausage and pork

材料 Ingredients

蘿蔔 4 斤
粘米粉半斤（2 1/2 杯）
粟粉 3 湯匙
臘腸 2 條
臘肉半條
蝦米 2 雨
芫茜、葱各適量
乾葱 2 粒
冰片糖 1/4 片
2.4 kg white radishes
300 g long-grain rice flour (2 1/2 cups)
3 tbsp cornstarch
2 Chinese preserved pork sausages
1/2 piece Chinese preserved pork belly
75 g dried shrimps
coriander
spring onion
2 shallots
1/4 red sugar slab

調味料 Seasoning

鹽 2 1/2 茶匙
糖半茶匙
雞粉半茶匙
魚露 1 1/2 湯匙
胡椒粉少許
2 1/2 tsp salt
1/2 tsp sugar
1/2 tsp chicken bouillon powder
1 1/2 tbsp fish sauce
ground white pepper

做法 Method

1. 蘿蔔刨皮，刨絲或切條；臘腸及臘肉切粒；蝦米洗淨，切粒。

2. 芫茜及葱切碎；乾葱拍碎。

3. 粘米粉及粟粉篩勻，放於大盆內，下調味料拌勻，備用。

4. 下油 2 湯匙，加入乾葱碎爆香，取出，再下蝦米、臘腸及臘肉炒勻，盛起。

5. 放入蘿蔔絲拌炒煮片刻，待滾後下冰片糖 1/4 片及清水適量拌勻，加入 3/4 份量臘味拌勻，倒入粘米粉內，快手攪拌至略稠。

6. 糕盆塗上油，倒入蘿蔔粉料，略按，鋪上餘下之臘味，隔水中大火蒸 50 分鐘，最後以芫茜及葱裝飾即可。

1. Peel the radishes. Grate into fine shreds or cut into strips. Dice the sausages and preserved pork belly. Rinse the dried shrimps and dice them.

2. Finely chop the coriander and spring onion. Crush the shallots.

3. Sieve both rice flour and cornstarch into a big mixing bowl. Add seasoning and stir well.

4. Heat 2 tbsp of oil in a wok. Stir fry shallot until fragrant. Remove and discard. Put in dried shrimps, sausages and preserved pork belly. Stir well. Set aside.

5. In the same wok, put in the grated radishes. Stir and bring to the boil. Add 1/4 red sugar slab and some water. Stir again. Put in 3/4 of the ingredients from step 4. Mix well. Pour this mixture into the dry ingredients in the big mixing bowl. Stir quickly to mix well until slightly thicken.

6. Grease a steaming tray. Pour in the batter from step 5 and slightly press. Arrange the remaining ingredients from step 4 on the surface. Steam over medium-high heat for 50 minutes. Garnish with coriander and spring onion at last. Serve.

 小煮意

- 選重身的蘿蔔,多汁美味,是炮製蘿蔔糕的最佳之選。

- 煮蘿蔔時,清水的份量需要自行判斷,因切條的蘿蔔汁較少;刨絲的蘿蔔汁會較多。若想蘿蔔糕質感稠身、喜歡煎吃的話,建議少放清水炒煮。

- 冰片糖的色澤比片糖淺,蒸出來的蘿蔔糕較白。

- Pick radishes that feel heavy in your hand. This is a sign that the radish is juicy and fresh. The radish cake would taste better if the radish is good.

- When you cook the shredded radish, you have to judge the water content according to your experience. Radish that is cut into strips tends to give less water. Grated radish, on the other hand, tends to give more water. If you like your radish cake firmer and denser, or if you always pan-fry your radish cake, it's advisable to add little water.

- Red sugar slab is lighter in colour than raw cane sugar slab. The radish cake will appear whiter after steamed.

Deep-fried mini twisted pretzels

五香蛋散仔

材料 Ingredients

麵粉半斤	300 g flour
雞蛋 2 個（大）	2 large eggs
南乳 2 湯匙	2 tbsp fermented tarocurd
五香粉 2 茶匙	2 tsp five-spice powder
豬油 2 湯匙	2 tbsp lard
白芝麻少許	white sesames
糖 1 茶匙	1 tsp sugar
鹽 3/4 茶匙	3/4 tsp salt
水 3 湯匙	3 tbsp water

做法 Method

1. 麵粉放於桌面，中間開成窩形，加入雞蛋、鹽、南乳、五香粉、糖、豬油及水拌勻，再慢慢拌入麵粉，搓成軟滑麵糰，用濕布蓋着。

2. 桌面灑上乾麵粉，放上麵糰用擀木棒壓成長薄塊狀，兩面灑上芝麻，切成 1 吋 x 2 吋的小長條，在每塊小長條的中間位置剞一刀，將末端穿過去。

3. 燒熱半鍋油（剛煮沸即可），放入小長條以小火炸至微黃色，用隔篩盛起，吸去多餘油分，待涼後享用，或置於玻璃瓶儲存。

1. Pour flour on a counter and make a well at the centre. Put eggs, salt, fermented tarocurd, five-spice powder, sugar, lard and water in the well. Mix the ingredients in the well first. Then push in the flour a little at a time. Knead into smooth dough. Cover with a damp towel.

2. Flour your counter. Put the dough on. Roll it out thinly into an elongated shape with a rolling pin. Sprinkle sesames on both sides. Cut into strips about 1 by 2 inches. Make a cut at the centre along the length. Put one end of the dough strip through the hole.

3. Heat 1/2 pot of oil until hot. Put in the twisted dough strips and deep fry them over low heat until golden. Remove from oil with a strainer ladle. Leave them on paper towel to absorb excess oil. Let cool and serve. Or, store in airtight glass container for later consumption.

省卻搓粉的步驟，可用市售的雲吞皮代替，但自己搓揉麵糰可加入五香粉及南乳，滋味倍增。

If you don't feel like making your own dough from scratch, you may use frozen wanton skin straight. But you can only add five-spice powder and fermented tarocurd to the mini pretzels only if you make your own dough. They taste a whole lot better.

芝
麻
蝦 Sesame shrimp toast

材料 Ingredients

蝦肉 12 兩
肥豬肉少許（煮熟、切幼粒）
方包 8 片
白芝麻半杯
蛋白 2 湯匙
芫茜葉數片
450 g shelled shrimps
fatty pork (boiled in water until
done, finely diced)
8 slices sandwich bread
1/2 cup white sesames
2 tbsp egg whites
coriander leaves

調味料 Seasoning

麻油 3/4 茶匙
鹽 3/4 茶匙
粟粉 2 湯匙
胡椒粉少許
3/4 tsp sesame oil
3/4 tsp salt
2 tbsp cornstarch
ground white pepper

做法 Method

1. 蝦肉用生粉及鹽各 1 茶匙醃 5 分鐘，洗淨，用乾布徹底壓乾水分，拍成蝦茸，再用刀背略剁。

2. 蝦肉放入碗內，下調味料、蛋白及肥豬肉順一方向拌勻，見略帶黏性，再撻至起膠，冷藏半小時。

3. 將每塊方包切成 6 小件，蝦膠塗抹包面，黏滿白芝麻，放上芫茜葉，放於油鑊用小火炸至金黃色，趁熱享用。

1. Marinate the shrimps with 1 tsp of caltrop starch and 1 tsp of salt. Leave them for 5 minutes. Rinse well. Wrap them in a dry towel and squeeze them to wipe dry. Crush them with the flat side of a knife. Then chop them up with the back of a knife.

2. Put the minced shrimp into a bowl. Add seasoning, egg whites and fatty pork. Stir in one direction until sticky. Lift the minced shrimp mixture and slap it hard onto the chopping board. Refrigerate for 30 minutes.

3. Cut each slice of bread into 6 pieces. Spread some minced shrimp mixture on each piece. Coat the minced shrimp with white sesames. Place the coriander leave on top. Heat oil in a wok and deep fry the shrimp toast over low heat until golden. Serve hot.

小煮意

將方包先放入雪櫃冷藏片刻，炸後更鬆脆。

Refrigerate the bread briefly before use. The toast tastes even crispier that way.

雞絲粉皮

Ribbon glass noodles with shredded chicken

 小煮意

鮮粉皮於南貨店有售，比乾貨的質感較佳。

You can get fresh glass noodles from Shanghainese grocery stores. They taste better than dried ones.

材料 Ingredients

嫩雞半隻
鮮粉皮 4 張
韭黃 2 兩
青瓜絲及甘筍絲各少許
銀芽 4 兩
紅椒 1 隻（切絲）
薑 2 片
蔥 2 條（切段）
白芝麻 2 湯匙

1/2 young chicken
4 sheets fresh glass noodles
75 g yellow chives
shredded cucumber
shredded carrot
150 g mung bean sprouts
1 red chilli (shredded)
2 slices ginger
2 sprigs spring onion (cut into
short lengths)
2 tbsp white sesames

調味料 Seasoning

生抽 1 1/2 湯匙
辣油半茶匙
芝麻醬 2 湯匙
麻油 1 湯匙
糖 1/4 茶匙
熟油 2 湯匙

1 1/2 tbsp light soy sauce
1/2 tsp chilli oil
2 tbsp sesame paste
1 tbsp sesame oil
1/4 tsp sugar
2 tbsp cooked oil

做法 Method

1. 雞洗淨，抹乾水分，抹上鹽、胡椒粉及紹酒各少許，放入蔥段及薑片，隔水蒸熟，撕成雞絲，備用。

2. 鮮粉皮抹淨，切粗條，與生抽 1 1/2 湯匙及少許麻油拌勻，放於碟上。

3. 鋪上雞絲，放上已炒熟的銀芽、青瓜絲、甘筍絲及韭黃，下調味料拌勻，最後灑上芝麻及紅椒絲，凍吃較佳。

1. Rinse the chicken and wipe dry. Rub salt, ground white pepper and Shaoxing wine over the chicken. Put spring onion and ginger into the chicken. Steam until done. Let cool and tear into shreds.

2. Wipe the glass noodles clean. Cut into thick ribbons. Toss with 1 1/2 tbsp of light soy sauce and a dash of sesame oil. Save on a serving plate.

3. Arrange the shredded chicken over the noodles. Top with fried mung bean sprouts, cucumber, carrot and yellow chives. Add seasoning and mix well. Sprinkle with sesames and red chilli at last. Serve chilled is the best.

咖喱角 Samosa

材料 Ingredients

牛肉半斤
洋葱 1 個（小）
乾葱 3 粒
馬鈴薯 1 個
方型薄餅皮半包
麵粉 3 湯匙
咖喱粉 1 湯匙

300 g beef
1 small onion
3 shallots
1 potato
1/2 pack square pizza base
3 tbsp flour
1 tbsp curry powder

醃料 Marinade

鹽半茶匙
生抽 1 茶匙
胡椒粉 1/4 茶匙
水 1 湯匙

1/2 tsp salt
1 tsp light soy sauce
1/4 tsp ground white pepper
1 tbsp water

做法 Method

1. 牛肉洗淨，剁爛，下醃料拌勻；洋葱切粒；乾葱拍成茸。
2. 馬鈴薯刨去皮，切塊，放入滾水煮至軟身，壓成薯茸。
3. 燒熱油 3 湯匙，下乾葱茸、洋葱及咖喱粉爆香，加入牛肉炒熟，拌入薯茸及麵粉炒至水分收乾，盛起待涼，成為餡料。
4. 將薄餅皮裁剪成 6.5cm x 12.5cm（見下圖），包入餡料依序沿虛線摺成三角形，以麵粉漿黏着餅皮邊，收口，放入油鑊炸至金黃色即可。

1. Rinse the beef and finely chop it. Add marinade and mix well. Dice onion. Crush the shallots and chop finely.

2. Peel the potato and cut into pieces. Boil in water until soft. Mash finely.

3. Heat 3 tbsp of oil in a wok. Stir fry shallot, onion and curry powder until fragrant. Add beef and stir until done. Add mashed potato and flour. Stir until it dries up. Set aside to let cool. This is the filling.

4. Cut the pizza base into desired size (see as below). Wrap some filling on it and pinch into a triangle according to the sequence. Seal the seam with flour slurry. Pinch well. Deep fry in hot oil until golden. Serve.

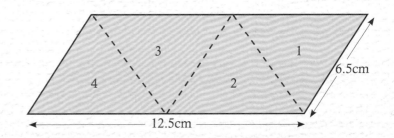

小煮意

- 包好後，宜將收口面向下放好；煎時也建議將收口面向下，可緊黏不易張開。
- 餡料可悉隨尊便，雞肉、牛肉、素肉或其他肉類皆可。

- After you wrap the filling in the samosa, put it down on the counter with the seam facing down. Similarly, when you fry it, you should put it with the seam facing down. That would ensure the seam won't break.

- You may use any meat you like in the filling, such as chicken or beef. Even vegetarian meat is fine.

Bacon and chive pancakes

材料 Ingredients

煙肉或火腿 4 安士
韭菜 4 兩
雞蛋 2 個
麵粉 8 安士（量杯計）
糯米粉 2 湯匙
水 10 安士
112 g smoked bacon or ham
112 g chives
2 eggs
235 ml flour (measured in cup)
2 tbsp glutinous rice flour
295 ml water

調味料 Seasoning

鹽 1 茶匙
糖半茶匙
雞粉半茶匙
胡椒粉少許
1 tsp salt
1/2 tsp sugar
1/2 tsp chicken bouillon powder
ground white pepper

小煮意

想變化口味，可用海鮮代替肉類。

For a variation, use seafood instead of bacon or ham.

做法 Method

1. 麵粉及糯米粉篩勻，加入水拌勻，拌入調味料及雞蛋，備用。
2. 煙肉切幼粒；韭菜洗淨，切粒，全部放入麵粉漿內拌勻。
3. 燒熱平底鑊，下少許油，舀入一杓麵粉漿，用慢火煎至兩面金黃色即可。

1. Sieve flour and glutinous rice flour together into a mixing bowl. Add water and stir well. Put in seasoning and eggs. Stir again.

2. Dice bacon finely. Rinse chives and dice them. Put both bacon and chives into the batter from step 1. Stir well.

3. Heat a pan and put in some oil. Pour a ladle of the batter in the pan. Fry over low heat until both sides golden. Serve.

鹹·小吃

Soy-marinated
kohlrabi

豉油浸沖菜

生抽冰糖汁待涼後才醃製沖菜，否則欠爽脆口感。

Make sure you leave the soy marinade to cool completely before you put in the kohlrabi. Otherwise, it'd lose its crunch.

材料 Ingredients

沖菜 1 斤
指天椒 5 隻
生抽 12 安士
冰糖 10 至 12 兩
麻油少許
600 g kohlrabi
5 bird's eye chillies
350 ml light soy sauce
375 g to 450 g rock sugar
sesame oil

做法 Method

1. 沖菜洗掉砂粒，切片或切條，用凍滾水浸約 1 小時，擠乾水分，備用。
2. 將生抽及冰糖用慢火煮至糖溶化，加入指天椒，待涼。
3. 將沖菜、調味汁及麻油放入玻璃瓶，浸泡 1 至 2 天即可食用，或儲存於雪櫃也可。

1. Rinse the kohlrabi well to remove any sand. Slice it or cut into strips. Soak in cold drinking water for 1 hour. Squeeze dry.

2. In a pot, cook light soy sauce and rock sugar over low heat until sugar dissolves. Put in bird's eye chillies. Let cool.

3. Put kohlrabi, soy marinade and sesame oil into a glass bottle with lid. Let the flavours mingle for 1 to 2 days and serve. You may also keep it in the fridge.

錦滷雲吞

Deep-fried wontons in sweet and sour sauce

材料 Ingredients

青、紅甜椒各 1 個（切片）
洋葱 1 個（切片）
罐裝蘑菇 4 安士（切片）
雲吞皮 2 兩
1 green bell pepper (sliced)
1 red bell pepper (sliced)
1 onion (sliced)
112 g canned button mushrooms (sliced)
75 g wanton skin

雲吞材料 Wonton filling

蝦肉 4 兩
熟肥豬肉 2 湯匙
冬菇 2 朵
150 g shelled shrimps
2 tbsp cooked fatty pork
2 dried shiitake mushrooms

雲吞調味料 Wonton seasoning

蛋黃半個
鹽、油及胡椒粉各少許
1/2 egg yolk
salt
oil
ground white pepper

汁料 Sweet and sour sauce

番茄醬 4 湯匙
白醋 3 湯匙
鹽半茶匙
糖 5 湯匙
水 6 安士
4 tbsp ketchup
3 tbsp white vinegar
1/2 tsp salt
5 tbsp sugar
175 ml water

生粉水 Caltrop starch slurry

生粉 3 茶匙
水 2 安士
3 tsp caltrop starch
60 ml water

做法 Method

1. 蝦肉剁爛；熟肥豬肉切粒；冬菇用水浸軟，去蒂，蒸熟後切粒。
2. 將雲吞材料拌勻，用雲吞皮包好，放入熱油內炸至金黃色，瀝乾油分。
3. 燒熱少許油，下洋葱炒香，加入青、紅甜椒及蘑菇炒勻，盛起。
4. 煮滾汁料，煮至濃稠時，加入上述洋葱料拌炒，下生粉水埋獻，與炸雲吞一拼上桌享用。

1. Finely chop the shrimps. Dice the fatty pork. Soak shiitake mushroom in water until soft. Remove the stems. Steam shiitake mushrooms until done. Dice them.

2. Mix all wanton filling ingredients together. Wrap some filling in each piece of wanton skin. Deep fry in hot oil until golden. Drain well.

3. Heat some oil in a wok. Stir fry onion until fragrant. Add bell pepper and button mushrooms. Stir well. Set aside.

4. Boil the sweet and sour sauce and cook till it thickens. Put in the onion mixture from step 3 and toss well. Stir in caltrop starch slurry. Serve on the side with deep-fried wantons.

≪ 小煮意 ≫

包餡料時份量不要太多，炸出來的雲吞才香脆可口。

Do not wrap too much filling in the wantons. Otherwise, they won't be crispy after fried.

Veggie and seafood pancakes 蔬菜海鮮煎餅

材料 Ingredients

四季豆 5 條
洋葱 1/4 個
秀珍菇半碗
甘筍 1/4 個
紫椰菜 1/4 個
黃甜椒 1/4 個
紅甜椒 1/4 個
蝦肉 2 兩
魷魚 2 兩
5 pods French beans
1/4 onion
1/2 bowl oyster mushrooms
1/4 carrot
1/4 purple cabbage
1/4 yellow bell pepper
1/4 red bell pepper
75 g shelled shrimps
75 g squids

麵糊料 Batter

麵粉 6 安士（量杯計）
粘米粉 2 湯匙
雞蛋 2 個
水 8 安士
油 1 湯匙
鹽 1/4 茶匙
175 ml flour (measured in cup)
2 tbsp long-grain rice flour
2 eggs
235 ml water
1 tbsp oil
1/4 tsp salt

調味料 Seasoning

鹽半茶匙
糖 1/4 茶匙
雞粉及胡椒粉各少許
1/2 tsp salt
1/4 tsp sugar
chicken bouillon powder
ground white pepper

做法 Method

1. 麵粉及粘米粉篩勻，下鹽及水拌勻，雞蛋分數次加入攪勻，待 10 分鐘。下油待 10 分鐘後拌勻。

2. 所有蔬菜及秀珍菇洗淨，抹乾，切條。

3. 蝦肉用生粉及鹽各半茶匙醃 5 分鐘，洗淨及抹乾，切粒；魷魚洗淨，切粒。

4. 熱鑊下少許油，加入洋蔥爆香，放入所有材料炒香，下調味料拌炒，上碟，待涼後加入麵糊內。

5. 燒熱平底鑊，下油至熱，舀入一杓蔬菜麵糊，用慢火煎至兩面金黃色及熟透即可。

1. Sieve the flour and rice flour together. Add water and salt. Mix well. Add a little whisked egg at a time and mix well after each addition. Let the batter sit for 10 minutes. Add oil. Wait for 10 minutes. Stir well.

2. Rinse all vegetables and oyster mushrooms. Wipe dry. Cut into strips.

3. Mix the shrimps with 1/2 tsp of caltrop starch and 1/2 tsp of salt. Leave them for 5 minutes. Rinse well. Wipe dry and dice them. Rinse the squid and dice it.

4. Heat oil in a wok. Stir fry onion until fragrant. Put in all ingredients and stir well. Add seasoning and stir further. Set aside and let cool. Stir them into the batter from step 1.

5. Heat a pan and add some oil. Ladle the batter into the pan. Fry over low heat until both sides golden and cooked through. Serve.

小煮意

可發揮創意，配搭不同的海鮮、煙肉、火腿或腸類。

You may use other seafood of your choice. Or try bacon, ham or sausages.

五香茶葉蛋 Tea-marinated eggs

- 調味汁加入冬菇蒂，味道更香濃惹味。
- 先將蛋殼輕輕敲裂，再浸 1 小時，使調味汁容易滲入雞蛋內。
- Shiitake mushroom stems add an extra depth and richness to the marinade.
- Cracking the egg shells and leaving the eggs in the marinade for 1 hour help the flavours to infuse into the eggs.

材料 Ingredients

雞蛋 10 個
紅茶葉 1 湯匙
鐵觀音茶葉 1 湯匙
冬菇蒂 1 兩
10 eggs
1 tbsp black tea leaves
1 tbsp Iron Buddha tea leaves
38 g stems of dried shiitake mushrooms

調味料 Seasoning

八角 2 粒
花椒半茶匙
糖 1 湯匙
生抽 1 湯匙
老抽 2 湯匙
鹽 2 茶匙
清水 5 杯
2 cloves star anise
1/2 tsp Sichuan peppercorns
1 tbsp sugar
1 tbsp light soy sauce
2 tbsp dark soy sauce
2 tsp salt
5 cups water

做法 Method

1. 將茶葉、冬菇蒂及調味料放入深鍋內煮滾，以慢火煮約 10 分鐘。
2. 雞蛋放入冷水內，煮約 10 分鐘至熟，盛起。
3. 輕輕敲裂蛋殼，放入茶葉調味汁內，以慢火煮約 15 分鐘，熄火，浸泡 1 小時，即可享用。

1. Put the tea leaves, mushroom stems and seasoning into a deep pot. Bring to the boil. Then turn to low heat and simmer for 10 minutes.
2. Put the eggs into a pot of cold water. Cook over high heat for 10 minutes until hard-boiled. Drain.
3. Gently tap the eggs on the counter to crack the shells. Put the eggs into the marinade from step 1. Cook over low heat for 15 minutes. Turn off the heat and let them soak in the marinade for 1 hour. Serve.

暹邏脆香魚

Deep-fried bonito flakes in Thai dressing

材料 Ingredients

柴魚 3 兩
洋葱 1 個
乾葱 4 粒
紅椒 1 隻
炸花生 2 兩
113 g dried bonito
1 onion
4 shallots
1 red chilli
75 g deep-fried peanuts

調味料 Seasoning

青檸汁 2 茶匙
鹽半茶匙
糖 2 湯匙
麻油及胡椒粉各少許
水 3 湯匙
2 tsp lime juice
1/2 tsp salt
2 tbsp sugar
sesame oil
ground white pepper
3 tbsp water

做法 Method

1. 柴魚用水浸軟，撕碎，瀝乾水分，放入油鑊內炸脆，瀝乾油分。
2. 洋葱、乾葱及紅椒洗淨，切絲。
3. 燒熱少許油，加入乾葱、洋葱及紅椒爆香，下脆柴魚炒勻，拌入調味料煮勻，潷酒，最後下炸香花生，上碟，灑上麻油享用。

1. Soak the bonito in water until soft. Tear into shreds. Drain well. Deep fry in oil until crispy. Drain again.

2. Rinse onion, shallots and red chilli. Shred them.

3. Heat oil in a wok. Stir fry shallot, onion and red chilli until fragrant. Put in the bonito. Toss well. Stir in seasoning and cook briefly. Sizzle with wine. Lastly, top with deep-fried peanuts. Save on a serving plate. Drizzle with sesame oil. Serve.

❧ 小煮意 ❧

柴魚又名鰹魚乾，是加工製成的魚乾，也可用其他小魚代替。

Dried bonito, or Katsuo in Japanese, is a dried fish in the mackerel family. You may use other small dried fish instead.

臘味芋頭糕

材料 Ingredients

芋頭 1 個（約 1 斤）
粘米粉半斤（2 1/2 杯）
蝦米 1 兩
臘腸 2 條
臘肉半條
水 1 1/4 杯
乾葱 4 粒

1 taro (about 600 g)
300 g (2 1/2 cups) long-grain rice flour
38 g dried shrimps
2 Chinese preserved pork sausages
1/2 piece Chinese preserved pork belly
1 1/4 cup water
4 shallots

調味料 Seasoning

生抽 1 湯匙
鹽 2 茶匙
糖 1 茶匙
胡椒粉少許
五香粉 2 茶匙
水 6 安士

1 tbsp light soy sauce
2 tsp salt
1 tsp sugar
ground white pepper
2 tsp five-spice powder
175 ml water

做法 Method

1. 芋頭切粒，隔水蒸 15 分鐘。
2. 臘腸及臘肉切粒；蝦米洗淨，切粒。
3. 燒熱油爆香乾葱，棄去，加入蝦米爆炒，再下臘味炒片刻，盛起。
4. 原鑊下芋粒爆炒，加入調味料炒勻，盛起。
5. 粘米粉用 1 1/4 杯水調勻，放入芋粒及 2/3 份量之臘味，拌勻。
6. 糕盆塗上油，倒入臘味粉漿，放入餘下之臘味鋪面，用大火蒸 1 小時即可。

1. Dice the taro. Steam for 15 minutes.

2. Dice the preserved sausages and pork belly. Rinse dried shrimps. Dice them.

3. Heat some oil in a wok and stir fry the shallots until fragrant. Remove and discard. Put in dried shrimps and stir briefly. Then add preserved sausages and pork belly. Stir further. Set aside.

4. In the same wok, stir fry diced taro until fragrant. Add seasoning and stir well. Set aside.

5. In a mixing bowl, put in rice flour. Add 1 1/4 cup of water. Mix well. Put in the diced taro and 2/3 of the ingredients from step 3. Mix well.

6. Grease a steaming tray. Pour in the batter from step 5. Arrange the remaining ingredients from step 3 on the surface. Steam over high heat for 1 hour. Serve.

❧ 小煮意 ❧

宜選輕身的芋頭，口感粉糯。

Pick taro that feels light in your hand. That's a sign of starchiness.

Taro cake with preserved sausage and pork

炸腰果拼白飯魚乾

Deep-fried cashews with
dried whitebaits

小煮意

首先必須燒熱鑊，才加入凍油，毋須
待油滾起，凍油時下腰果，才能炸至
鬆脆的效果。

Make sure the wok is heated until red
hot before pouring in cold oil. You don't
need to wait till the oil heat up before
you put in the cashews. This is the
secret trick to their crispiness after fried.

材料 Ingredients

腰果 4 兩
白飯魚乾 2 兩
淮鹽半茶匙
油 3 杯
150 g cashews
75 g dried whitebaits
1/2 tsp five-spice salt
3 cups oil

做法 Method

1. 腰果放入滾水內，下鹽 1 茶匙煲約 5 分鐘，隔
   去水分，吹乾或置陽光下曬至乾透。
2. 燒熱鑊，下凍油，加入腰果以慢火炸至微黃色，
   期間不停移動腰果，最後調大火略炸，上碟，
   待涼。
3. 白飯魚乾炸至金黃色及微乾，上碟待涼，與腰
   果一併灑上淮鹽享用。

1. Boil a pot of water and add 1 tsp of salt. Put
   in the cashews and cook for 5 minutes. Drain.
   Let them dry in the air or under the sun.

2. Heat a wok and pour in the cold oil. Put
   in the cashews and deep-fry over low heat
   until lightly browned. Jiggle the cashews
   continuously throughout the process. Turn up
   the heat at last and fry briefly. Drain and let
   cool.

3. Deep-fry the dried whitebaits until golden and
   crispy. Drain and let cool. Mix with cashews
   and sprinkle with five-spice salt. Serve.

醃酸子薑 Young ginger pickles

材料 Ingredients

嫩子薑 2 斤
酸梅 3 粒
鹽半湯匙
白米醋 1 斤
粗砂糖半斤
1.2 kg young ginger
3 sour plums
1/2 tbsp salt
600 ml rice vinegar
300 g coarse sugar

做法 Method

1. 白米醋及粗砂糖拌勻，煮至糖溶化，待涼備用。
2. 子薑洗淨，刮去皮，切薄片，灑入鹽醃 1 小時至子薑軟化（期間不斷翻動，令鹽能均勻醃漬子薑）。
3. 子薑取出，用凍開水沖去鹽分，放入已待涼的白米醋內，加入搗碎的酸梅醃數天即可。

1. Mix vinegar with sugar. Cook till sugar dissolves. Let cool.

2. Rinse the young ginger. Scrape off the skin with a metal spoon. Cut into thin slices. Sprinkle with salt. Let it steep in the brine for 1 hour until the ginger softens. (Stir the ginger from time to time so that the salt penetrates the ginger evenly.)

3. Take the ginger out and rinse in cold drinking water. Put it into the vinegar syrup that has cooled completely. Add grated sour plum. Leave it there for a few days. Serve.

小煮意

- 子薑一般盛產於初夏季節，大部份菜檔有售。
- 若想子薑帶粉紅色澤，可加入少許酸梅茸或酸梅醋同醃。
- Young ginger is available in early summer. You can get it from most greengrocers.
- If you want to add a pink tint to your young ginger, put in some grated sour plum or steep it in some plum vinegar.

沙薑鳳爪　Chicken feet in sand ginger sauce

材料 Ingredients

鳳爪 1 斤
600 g chicken feet

調味料 Seasoning

上湯 6 杯
沙薑粉 3 湯匙
紅辣椒 1 隻（切絲）
八角 2 粒
蒜頭 2 粒（切片）
鹽 3 湯匙
糖 1 茶匙
麻油 2 湯匙
胡椒粉少許
紹酒 2 湯匙

6 cups stock
3 tbsp ground sand ginger
1 red chillies (shredded)
2 cloves star anise
2 cloves garlic (sliced)
3 tbsp salt
1 tsp sugar
2 tbsp sesame oil
ground white pepper
2 tbsp Shaoxing wine

做法 Method

1. 鳳爪洗淨，用滾水煮約 5 分鐘，過冷河，備用。

2. 煮滾調味料，放入鳳爪煲約 10 分鐘，再浸 1 小時，上碟，灑上麻油即可食用。

1. Rinse chicken feet and blanch them in boiling water for 5 minutes. Rinse with cold water. Set aside.

2. Boil the seasoning. Put the chicken feet in and cook for 10 minutes. Turn off the heat and leave them in the seasoning for 1 hour. Transfer onto a serving plate. Drizzle with sesame oil. Serve.

❧ 小煮意 ❧

鳳爪略灼後過冷河，皮質爽滑，嚼感佳！

After blanching the chicken feet, make sure you rinse them in cold water until completely cooled. The skin would taste more springy that way.

焗腸仔卷

Sausage rolls

材料 Ingredients

脆皮腸仔 12 條
雞蛋 1 個
酥皮 1 包

12 sausages with crispy skin
1 egg
1 pack frozen puff pastry

做法 Method

1. 酥皮解凍，放於室溫待略軟。
2. 將酥皮擀成 1/4 吋厚之長方形，用刀切成半吋濶長條。
3. 每塊小酥皮包入腸仔一條，捲起，於摺口位置塗上少許蛋汁收口，放於灑上麵粉的焗盆內。
4. 將焗爐調至 220℃，預熱 5 分鐘。放入腸仔卷焗 15 分鐘，見外皮金黃色，掃上蛋汁再焗 2 分鐘即可。

1. Thaw the puff pastry. Leave it at room temperature to warm it up a little.

2. Roll the puff pastry into a rectangle about 1/4 inch thick. Cut into strips about 1/2 inch thick with a knife.

3. Wrap a sausage in each strip of puff pastry. Roll it up. Seal the seam by brushing egg wash on it. Put them onto a baking tray with flour sprinkled on.

4. Preheat an oven to 220°C for 5 minutes. Bake the sausage rolls for 15 minutes until golden. Brush egg wash on top and bake 2 more minutes. Serve.

❧ 小煮意 ❧

現成的酥皮較為方便，毋須經過多重搓揉摺疊的步驟，於大型超級市場有售。

Ready-made puff pastry is a convenient alternative to making your own from scratch. You don't need to fold and stack the dough repeatedly. You can get frozen puff pastry from most large-scale supermarkets.

The four coagulating agents

凝固四主角

軟滑Q彈的甜品凍糕，一直是人氣甜吃。如何有效地凝固成軟糕，坊間有多種凝固作用的食用素材，各有不同特點。

蒟蒻凍粉 Konjac powder

取自蒟蒻芋的地下塊莖加工製成，味道清淡，完成的凍糕口感彈牙、煙韌，有清爽的感覺，個人較喜愛使用此款作凝固之用。

Made from the underground tuber of konjac plant, it tastes rather bland. Jelly made with konjac powder has a chewy and springy texture. This is my favourite gelling agent for chilled desserts.

魚膠粉 Gelatine powder

與魚膠片相似，只是為粉狀形式，浸水後用熱水座溶，較方便使用，而且經濟實惠，大多數人選用，但製成的軟糕較軟睑，欠彈牙質感，溶解時也容易起粒。

Similar to gelatine leaves, it is only a variation of physical form. Soak it in water and then heat it over a pot of simmering water to dissolve it. It's more convenient than using gelatine leaves and it's considerably cheaper. Most people make jelly with gelatine powder, but the desserts made with it do not have the same springy mouthfeel. The desserts may also turn lumpy when they melt.

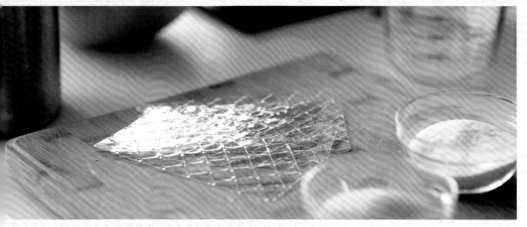

魚膠片 Gelatine leaves

由動物骨骼提煉的膠質原料，帶腥味，使用時用水浸軟，擠乾後用熱水座溶或煮溶，一般用於西式甜品，凝固後質感較硬。

Made from animal bones and tendons, gelatine tastes gamey. You have to soak them in water until soft. Then squeeze dry and heat them over a pot of simmering water or cook them until they dissolves. They are used mostly in western jelly and puddings. Desserts made with gelatine leaves tend to be firmer in texture.

大菜 Agar-agar

由海藻提煉出來的植物性膠質原料，由於價錢廉宜，一直以來廣受使用，最常製成大菜糕或啫喱等，口感爽脆。

It's the botanical protein extracted from seaweed. As it's very frugal in price, it's a very popular ingredient among the general public. Agar-agar is made into jelly and it has a softer texture.

Pomelo konjac jelly

〜 小煮意 〜

以蒟蒻凍粉製作糕點，比用魚膠粉或大菜，質感更彈牙。

Konjac has a different texture from gelatine or agar-agar. Konjac is chewy but less jiggly than gelatine.

材料 Ingredients

蒟蒻凍粉 4 1/2 湯匙
糖 4 至 6 湯匙
柚子蜜 2 湯匙
清水 16 安士
4 1/2 tbsp konjac powder
4 to 6 tbsp sugar
2 tbsp Korean pomelo honey
470 ml water

做法 Method

1. 蒟蒻凍粉及糖拌勻，注入清水開火不停攪拌，煮至砂糖溶化，加入柚子蜜拌勻。
2. 傾入容器內，放入雪櫃冷凍，清甜可口。

1. Mix konjac powder with sugar in a pot. Add water and turn on the heat. Keep stirring until sugar dissolves. Stir in pomelo honey.

2. Pour into jelly cups or container of your choice. Refrigerate until set. Serve.

椰皇燉鮮奶蛋白

Double-steamed egg white milk custard in coconut

材料 Ingredients

椰皇 2 個
椰皇水 7 安士（量杯計）
蛋白 4 個（約 150 克）
糖 2 湯匙
鮮奶 7 安士
白米醋 1/4 茶匙
2 smoked coconuts
205 ml coconut water
4 egg whites (about 150 g)
2 tbsp sugar
205 ml milk
1/4 tsp rice vinegar

做法 Method

1. 用刀破開椰皇頂部，倒出椰皇水留用。

2. 將熱水倒入椰皇內，傾出，座於碗內以錫紙固定位置。

3. 鮮奶及糖用小火煮至微熱，待涼備用。

4. 蛋白、白米醋及椰皇水拂勻，倒入鮮奶徐徐拌勻，用密篩過濾，注入量杯內。

5. 椰皇內倒入鮮奶蛋白，以錫紙或紗紙覆蓋，用中火蒸約 20 分鐘，熄火，再焗 5 分鐘，熱吃或冷吃皆可。

1. Cut off the top of the smoked coconuts. Drain the coconut water for later use.

2. Pour hot water into the hollow smoked coconuts. Drain. Put them into a bowl and keep them upright by putting some scrunched aluminium foil under them.

3. In a pot, cook milk and sugar over low heat until warm. Set aside to let cool.

4. Whisk the egg whites, rice vinegar and coconut water together until well incorporated. Pour in the milk. Pass the mixture through a sieve into a measuring cup.

5. Pour the egg white mixture into the hollow smoked coconuts. Cover the top with aluminium foil or mulberry paper. Steam over medium heat for 20 minutes. Turn off the heat and keep them in the steamer with the lid covered for 5 more minutes. Serve hot or chilled.

❧ 小煮意 ❧

- 要選重身及體型大的椰皇，水分充足。

- 將熱水倒入椰皇內，令椰殼預早受熱，加快燉製的時間。

- 加入白米醋，有凝固鮮奶及蛋白的作用。

- Pick smoked coconuts that are heavy and bulky. They tend to have more water inside.

- Pouring hot water into the smoked coconut and draining help pre-heat them. That would shorten the steaming time.

- Rice vinegar helps the egg white mixture set.

紅棗雪耳凍糕

Red date and white fungus jelly

材料 Ingredients

無核紅棗 2 兩
雪耳 2 錢
蒟蒻凍粉 5 湯匙
糖 6 湯匙
清水 16 安士
75 g red dates (pitted)
8 g white fungus
5 tbsp konjac powder
6 tbsp sugar
470 ml water

做法 Method

1. 雪耳浸半小時，剪去硬蒂，洗淨。

2. 紅棗洗淨，用清水 16 安士浸片刻，放入攪拌機打成茸，隔出紅棗碎，將紅棗水倒回攪拌機內，放入雪耳打碎。

3. 將紅棗雪耳水放入煲內，加入糖及蒟蒻凍粉拌勻，不停攪拌，煮至砂糖溶化，熄火，放入容器內冷藏，凝固後即可享用。

1. Soak white fungus in water for 30 minutes. Cut off the roots. Rinse well.

2. Rinse red dates and soak them in 470 ml of water. Then pour them into a blender. Blend until fine. Pass the mixture through a sieve. Pour the red date juice back into the blender. Add white fungus and puree.

3. Pour the white fungus and red date mixture into a pot. Add sugar and konjac powder. Turn on the heat. Keep stirring until sugar dissolves. Turn off the heat and pour into jelly cups or container of your choice. Refrigerate until set. Serve.

小煮意

- 紅棗去核後，才放入攪拌機內打碎。
- 建議徹底隔去紅棗碎，只保留紅棗水，令凍糕軟滑可口。
- Make sure you pit the red dates before blending them in a blender.
- To make the jelly silky smooth, I prefer straining the red date juice to remove any gritty bits.

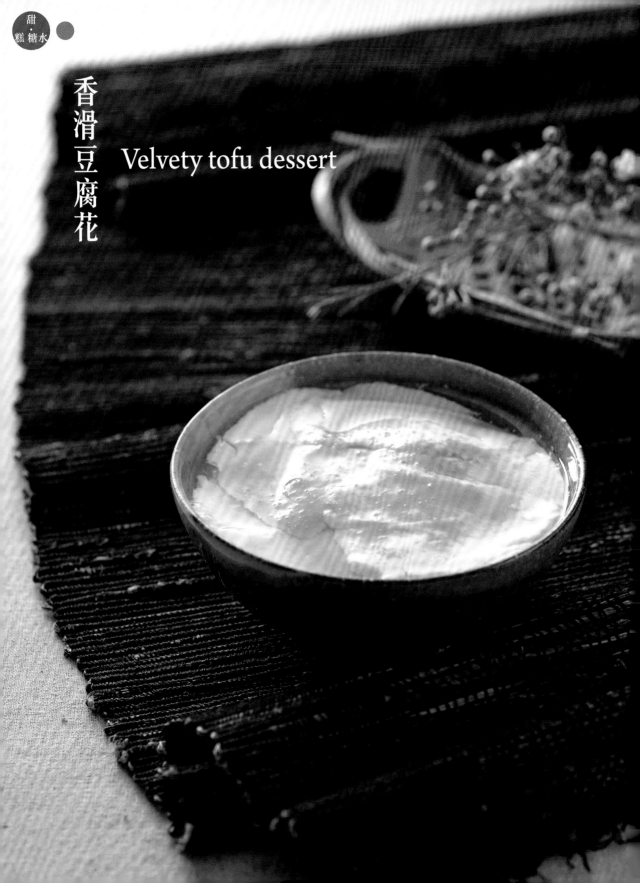

香滑豆腐花 Velvety tofu dessert

材料 Ingredients

黃豆半斤
清水 16 杯
食用石膏粉 2 茶匙
生粉 1 1/2 湯匙
薑汁 1 1/2 湯匙
凍滾水 4 安士
300 g soybeans
16 cups water
2 tsp food-grade gypsum
1 1/2 tbsp caltrop starch
1 1/2 tbsp ginger juice
120 ml cold drinking water

糖漿 Syrup

純正紅糖碎適量
清水 1 碗
crushed red sugar
1 bowl water

小煮意

- 食用石膏粉於藥材店或雜貨店有售，加入薑汁可去除豆腥味。

- 黃豆浸後去掉豆衣，令磨出來的豆漿更滑溜。

- You can get food-grade gypsum from Chinese herbal stores or grocery stores. Adding ginger juice helps remove the grassy taste of the soybeans.

- After soaking soybeans, removing their skin helps make the soymilk finer in texture.

做法 Method

1. 黃豆浸水 6 小時，洗去豆皮，加清水 16 杯放入攪拌機內，磨成豆漿，用紗布過濾，隔去豆渣，煮滾豆漿。

2. 食用石膏粉、生粉及薑汁，用凍滾水攪拌。

3. 將糖漿材料混和，煮至糖溶，備用。

4. 預備一個大容器，將石膏粉水及煮滾的豆漿依序分別撞入，不要攪動，加蓋，待 20 至 30 分鐘至凝固，舀出豆腐花，灑入糖漿品嘗。

1. Soak the soybeans in water for 6 hours. Drain. Rinse and remove the skin. Put soybeans and 16 cups of water into a blender. Blend into soymilk. Pass soymilk through cheese cloth to remove the dregs. Boil the soymilk.

2. Mix gypsum, caltrop starch and ginger juice with 1/2 bowl of cold drinking water. Mix well.

3. In a pot, put in syrup ingredients and cook until sugar dissolves.

4. In a large tub, pour in gypsum mixture from step 2 first and recently boiled soymilk. Do not stir it. Cover the lid and let it sit for 20 to 30 minutes until set. Put soft tofu into serving bowls. Drizzle with syrup. Serve.

Black sesame rolls 爽滑芝麻卷

材料 Ingredients

黑芝麻 4 兩	150 g black sesames
馬蹄粉 4 1/2 兩	165 g water chestnut starch
粟粉 2 湯匙	2 tbsp cornstarch
糖 10 安士	290 g sugar
清水 14 安士	415 ml water

做法 Method

1. 黑芝麻洗淨，瀝乾水分，用白鑊慢火炒香，放入攪拌機內，加水 6 安士磨成黑芝麻漿，用布袋過濾。

2. 馬蹄粉及粟粉用水 8 安士調勻，用密篩隔去粗粒。

3. 砂糖加入黑芝麻漿內煮溶，待至暖和，倒入馬蹄粉水徐徐攪拌。

4. 糕盆掃上油，倒入一碗黑芝麻粉漿，成為薄薄的一層，隔水蒸約 5 分鐘，取出待涼，輕輕捲成長條，切件享用。

1. Rinse black sesames and drain well. Toast them in a dry pan until fragrant. Blend black sesames with 175 ml of water until fine. Strain through cheese cloth or muslin.

2. Mix water chestnut starch and cornstarch with 235 ml of water. Pass through a sieve to remove any lumps.

3. In a pot, put in black sesame milk. Add sugar and turn on the heat. Cook until sugar dissolves and the mixture is warm. Slowly stir in water chestnut slurry from step 2.

4. Grease a steaming tray. Pour in 1 bowl of batter. Swirl to coat the bottom evenly and flatly. Steam for 5 minutes. Let cool. Roll it into a cylinder. Slice and serve.

椰汁鮮奶燉銀耳

Double-steamed white fungus in
coconut milk

材料 Ingredients

銀耳 (雪耳) 2 錢
紅棗 10 粒 (去核)
鮮奶 1 瓶 (235ml)
椰汁 1 罐 (200ml)
冰糖 3 兩
水 5 碗
8 g white fungus
10 red dates (pitted)
235 ml milk
200 ml coconut milk
113 g rock sugar
5 bowls water

做法 Method

1. 銀耳用水浸軟,剪去硬蒂,飛水,盛起,放入凍水內略洗,擠乾水分備用。

2. 冰糖放入鍋內,倒入水 5 碗煮至冰糖溶化,轉放入燉盅內。

3. 加入紅棗及銀耳,加蓋,燉約 1 小時,倒入鮮奶及椰汁,再燉 10 分鐘即成。

1. Soak the white fungus in water until soft. Cut off the roots. Blanch in boiling water. Drain. Rinse in cold water. Squeeze dry and set aside.

2. Put rock sugar into a pot. Add 5 bowls of water and cook till sugar dissolves. Transfer into double-steaming pot.

3. Put in red dates and white fungus. Cover the lid and double-steam for 1 hour. Add milk and coconut milk. Double-steam for 10 more minutes. Serve.

❧ 小煮意 ❧

若想節省時間,用慢火先煲煮紅棗、銀耳及冰糖 30 分鐘,再下鮮奶及椰汁拌勻,熄火即可。

To save cooking time, you don't have to double-steam the sweet soup. Just boil red dates, white fungus and rock sugar in water over low heat for 30 minutes first. Then add milk and coconut milk and stir well. Turn off the heat and serve.

楊枝甘露凍糕

Creamed mango, pomelo and sago jelly

材料 Ingredients

芒果 2 個
泰國柚子 2 塊
芒果味啫喱粉 2 盒
魚膠粉 2 湯匙
糖 3 湯匙
花奶 1 罐（380ml）
椰漿 1 罐（200ml）
滾水 2 杯

2 mangoes
2 slices Thai pomelo
2 boxes mango flavoured jelly powder
2 tbsp gelatine powder
3 tbsp sugar
1 can evaporated milk (380 ml)
1 small can coconut milk (200 ml)
2 cups boiling water

做法 Method

1. 芒果去皮，切幼粒；柚子去皮，撕成絲，備用。

2. 魚膠粉與凍滾水 4 湯匙拌溶。

3. 啫喱粉放入大碗內，下糖及滾水 2 杯攪勻至溶化，放入魚膠粉水、椰漿、花奶、柚子絲及芒果粒拌勻，盛起，冷藏至凝固即可。

1. Peel and core the mangoes. Dice the flesh finely. Peel the pomelo and break the pulp into individual juice sacs.

2. Mix gelatine with 4 tbsp of cold drinking water. Stir until it dissolves.

3. Put jelly powder into a large bowl. Add sugar and 2 cups of boiling water. Stir until sugar and jelly powder dissolve. Pour in the gelatine mixture, coconut milk, evaporated milk, pomelo pulp and diced mango. Mix well. Pour into containers of your choice. Refrigerate until set. Serve.

小煮意

泰國柚子多汁清甜，也可用沙田柚代替。

Thai pomelo tends to be juicier and sweeter. But you may also use regular pomelo from China instead.

香滑芝麻糊

Creamy black sesame sweet soup

材料 Ingredients

黑芝麻 4 兩
白米 5 湯匙
糖 12 安士
清水 10 杯
150 g black sesames
5 tbsp rice
340 g sugar
10 cups water

做法 Method

1. 黑芝麻洗淨，炒香備用；白米洗淨，用水浸透，備用。

2. 白米、黑芝麻及水 2 杯放入攪拌機內磨成黑芝麻漿，倒入布袋內過濾。

3. 煮滾清水 8 杯，下糖煮溶，水滾後倒入黑芝麻漿攪拌，待滾後加蓋，轉小火再滾片刻即成。

1. Rinse black sesames and toast them in a dry pan until fragrant. Rinse the rice and soak in water until soft. Set aside.

2. Put rice, black sesames and 2 cups of water into a blender. Blend into black sesame milk. Strain it through cheese cloth or muslin.

3. In a pot, boil 8 cups of water. Add sugar and cook till sugar dissolves. Bring to the boil. Pour in the black sesame milk. Stir well. Bring to the boil again and cover the lid. Turn to low heat and simmer briefly. Serve.

小煮意

- 黑芝麻磨成芝麻漿後，以布袋過濾，煮成的芝麻糊更幼滑。

- 黑芝麻用白鑊小火炒香，芝麻才不帶苦味。

- Straining the sesame milk through cheese cloth after blended helps make the sweet soup finer and smoother in texture.

- When you fry the sesames in a dry wok or pan, make sure you put it on low heat. The sweet soup may taste bitter if you burn the sesames.

黃金馬蹄糕

Golden water chestnut cake

材料 Ingredients

馬蹄半斤	300 g water chestnuts
馬蹄粉 8 安士（量杯計）	235 ml water chestnut starch (measured in cup)
菱粉 2 湯匙	2 tbsp water caltrop starch
吉士粉 5 湯匙	5 tbsp custard powder
澄麵 2 湯匙	2 tbsp wheat starch
糖 10 安士	290 g sugar
清水 30 安士	885 ml water

做法 Method

1. 馬蹄洗淨，去皮，切片或拍碎。
2. 馬蹄粉、菱粉、吉士粉及澄麵放入大碗內，加入清水 10 安士調勻，用密篩過濾雜質。
3. 餘下的 20 安士清水煲滾，下糖煮至溶。
4. 將粉漿略拌，徐徐加入 1/5 份量於糖水內，攪拌均勻，然後再將混合物倒入餘下的粉漿內，加入馬蹄粒拌勻。
5. 糕盆內塗上油，倒入馬蹄糊，用大火隔水蒸約 20 至 25 分鐘即可。

1. Rinse and peel water chestnuts. Slice or crush them with the flat side of a knife.

2. Put water chestnut starch, water caltrop starch, custard powder and wheat starch into a mixing bowl. Add 290 ml of water and mix well. Pass the batter through a sieve to remove any lumps.

3. Bring the remaining water to the boil. Add sugar and cook until it dissolves.

4. Stir to mix the batter well. Pour 1/5 of the batter into the syrup from step 3. Mix well. Then pour the mixture back into the remaining batter. Add water chestnuts. Mix again.

5. Grease a steaming tray. Pour in the batter. Steam over high heat for 20 to 25 minutes. Serve.

❧ 小煮意 ❧

- 馬蹄糕拌入菱粉，令糕品凝固稠身，雜貨店有售；若沒有菱粉，可用粟粉代替。

- 西式糕餅常用的吉士粉，使馬蹄糕帶香氣及黏稠，顏色亮麗，一般超市有售。

- Adding water caltrop starch to the batter helps make the cake firmer. You can get water caltrop starch from grocery stores. If you can't get water caltrop starch, use cornstarch instead.

- Custard powder is commonly used in Western-style custard and baked goods. It adds an eggy aroma and helps hold the cake together. It also gives the cake its golden colour. You can get it from most supermarkets.

蔗汁馬蹄糕

少變化，多口味

～ 小煮意 ～

鮮蔗汁也可用盒裝或樽裝代替。

You may use freshly squeezed sugar cane juice from herbal tea store. Or, you may get the boxed ones or bottles ones from supermarket.

Sugar cane and water chestnut cake

材料 Ingredients

馬蹄半斤
馬蹄粉 14 兩
鮮蔗汁 4 杯
粗砂糖 12 安士
清水 3 杯
300 g water chestnuts
525 g water chestnut starch
4 cups sugar cane juice
335 g coarse sugar
3 cups water

做法 Method

1. 馬蹄洗淨，去皮，切片或切粒，備用。
2. 馬蹄粉與水調勻，用密篩過濾雜質。
3. 將蔗汁、馬蹄粒及粗砂糖放入煲內，煮至糖溶後，熄火，倒入馬蹄粉水快手拌勻。
4. 糕盆內塗上油，倒入蔗汁馬蹄粉漿，隔水蒸 45 分鐘即可。

1. Rinse and peel water chestnuts. Slice or dice them.

2. Mix water with water chestnut starch. Pass the mixture through a sieve to remove any lump.

3. Put sugar cane juice, water chestnuts and sugar into a pot. Cook until sugar dissolves. Turn off the heat and pour in the water chestnut starch slurry from step 2. Stir quickly to combine.

4. Grease a steaming tray. Pour in the batter from step 3. Steam for 45 minutes. Serve.

眉豆馬蹄糕

Black-eyed bean and water chestnut layered cake

材料 Ingredients

眉豆 4 兩
馬蹄粉 8 安士（量杯計）
泰國薯粉 5 湯匙
片糖 10 兩
清水 5 1/2 杯
150 g black-eyed beans
235 ml water chestnut starch (measured in cup)
5 tbsp Thai tapioca starch
375 g raw cane sugar slab
5 1/2 cups water

◇ 小煮意 ◇

將馬蹄粉漿及眉豆粉漿相
間地加入蒸熟，若怕步驟
繁複，可將全部粉漿拌勻
後隔水蒸，但需要邊蒸邊
拌一會。

I divided the water chestnut
and black-eyed bean batter
into layers in this recipe.
If you think it's too much
work, you can always mix
all batter together and steam
it in one go. But you may
have to stir the batter in the
steaming process so that
the beans won't sink to the
bottom.

做法 Method

1. 眉豆洗淨，用水浸 2 至 3 小時，瀝乾水分，隔水蒸至豆粒軟脸，備用。

2. 馬蹄粉、薯粉及水 2 杯拌勻，用密篩過濾，備用。

3. 燒滾水 3 1/2 杯，加入片糖煮至溶，下馬蹄粉漿用慢火拌勻，熄火，快手攪拌，倒出粉漿 3 杯備用。

4. 餘下的粉漿加入眉豆，用慢火煮至半熟狀態，熄火，拌勻。

5. 糕盆內掃上油，先倒入 1 杯馬蹄粉漿蒸 5 分鐘，再加入眉豆粉漿蒸約 35 分鐘，最後加入餘下的 2 杯馬蹄粉漿，蒸 10 分鐘即可。

1. Rinse the black-eyed beans. Soak them in water for 2 to 3 hours. Drain. Steam until soft.

2. Mix water chestnut starch and tapioca starch with 2 cups of water. Pass the mixture through a sieve.

3. Boil 3 1/2 cups of water in a pot. Add sugar slab and cook until it dissolves. Add the water chestnut starch slurry from step 2. Cook over low heat and stir well. Turn off the heat. Stir quickly to combine. Set aside 3 cups of batter for later use.

4. Add black-eyed beans to the remaining batter. Cook over low heat until half-cooked. Turn off the heat and mix well.

5. Grease a steaming tray. Pour in 1 cup of water chestnut batter from step 3. Steam for 5 minutes. Top with a layer of black-eyed bean batter from step 4. Steam for 35 minutes. Lastly top with the remaining 2 cups of water chestnut batter. Steam for 10 more minutes. Serve.

做法 Method

1. 蛋白及糖拂打至企身,加入白米醋及鮮奶打勻,撇去泡沫,盛於碗內,用紗紙蓋面。
2. 燒滾水,架上蒸籠,排入碗以大火蒸7分鐘,熄火,待5分鐘即可享用。

1. Beat egg whites and sugar until foamy. Add rice vinegar and milk. Mix well. Skim off the bubbles. Pour into bowls. Cover with mulberry paper.

2. Boil water in a steamer. Put in the bowls and steam over high heat for 7 minutes. Turn off the heat and let the custard sit in the steamer with the lid covered for 5 more minutes. Serve.

香滑鮮奶燉蛋白

Double-steamed milk and egg white custard

材料 Ingredients

鮮奶 5 安士
蛋白 1 個（大）
糖約 1 茶匙
白米醋 5 滴

145 ml milk
1 egg white (large)
1 tsp sugar
5 drops rice vinegar

拂鮮奶蛋白時加入白米醋，令混合物更均勻，而且蒸後容易凝固。

Adding vinegar to the egg white helps the ingredients blend better when whisked. The custard also tends to set more easily.

椰汁香芒紫米

Black glutinous rice with coconut milk and mango

小煮意

黑糯米放入飯煲烹調，才能煮成濃稠粥狀。

You have to cook the black glutinous rice in a rice cooker to make it congee-like.

材料 Ingredients

黑糯米半斤	300 g black glutinous rice
芒果粒 1 杯	1 cup diced mango flesh
花奶 1 罐（380ml）	1 can evaporated milk (380 ml)
椰汁 8 安士	235 ml coconut milk
馬蹄粉 3 安士（量杯計）	85 ml water chestnut starch (measured in cup)
糖 12 安士	340 g sugar
水 14 安士	410 ml water
油 1 湯匙	1 tbsp oil

做法 Method

1. 黑糯米洗淨，用水浸 4 小時，瀝乾水分，放入飯煲內，下適量水煮成稠粥。
2. 水 10 安士調勻馬蹄粉，過濾，拌入椰汁及花奶，備用。
3. 煮滾水 4 安士，加入糖及油煮至糖溶化，撞入馬蹄粉漿煮至濃稠狀。
4. 黑糯米飯與馬蹄粉糖水拌勻，加入 3/4 份量芒果粒，盛於碗內，待涼至凝固，加入其餘芒果粒裝飾，倒入糖水、花奶及椰汁享用。

1. Rinse the black glutinous rice. Soak in water for 4 hours. Drain. Put into a rice cooker and water. Turn on the cooker and cook the rice into a thick congee.

2. Mix water chestnut starch with 295 ml of water. Pass it through a sieve. Stir in coconut milk and evaporated milk.

3. Boil the remaining water in a pot. Add sugar and oil. Cook till sugar dissolves. Pour in water chestnut slurry from step 2. Cook till it thickens.

4. Mix black glutinous rice congee with water chestnut batter from step 3. Stir well. Add 3/4 of the diced mango. Save in jelly cups or your preferred container. Let cool till set. Garnish with remaining diced mango. Pour some extra syrup, evaporated milk and coconut milk on top. Serve.

Almond and walnut sweet soup

杏仁合桃糊

❧ 小煮意 ❧

合桃及南杏攪拌前，要用白鑊慢火炒至微黃色，令香氣濃郁。

Before blending the walnuts and almonds, make sure you fry them
in a dry pan or wok until lightly golden. That would heighten the
nutty aroma of the dessert.

材料 Ingredients

南杏 5 錢
合桃仁 4 兩
白米 4 湯匙
冰糖 10 兩
杏仁油 1/8 茶匙
水 10 杯
19 g sweet almonds
150 g shelled walnuts
4 tbsp rice
375 g rock sugar
1/8 tsp almond essence
10 cups water

做法 Method

1. 白米洗淨,略浸 15 分鐘,備用。

2. 合桃仁及南杏放入白鑊內炒香,取出,與白米及水 1 1/2 杯放入攪拌機磨成漿,過濾備用。

3. 將杏仁合桃米漿放入鍋內,倒入水 10 杯及冰糖拌勻,煮至冰糖溶化及米漿濃稠,加入杏仁油,成為香滑滋養之杏仁合桃糊。

1. Rinse the rice and soak it in water for 15 minutes.

2. Stir fry walnuts and almonds in a dry pan until fragrant. Transfer rice, walnuts, almonds and 1 1/2 cups of water into a blender. Blend into walnut almond milk. Strain through a sieve.

3. Pour walnut almond milk into a pot. Add 10 cups of water and rock sugar. Turn on the heat and stir well. Cook till sugar dissolves and the walnut almond milk thickens. Keep stirring throughout the process. Add almond essence. Serve.

燕窩鮮奶燉萬壽果

Double-steamed papaya with bird's nest and milk

材料 Ingredients

夏威夷木瓜 2 個
燕窩 2 錢
鮮奶 8 安士
冰糖適量
水 2 杯
2 Hawaiian papayas
8 g bird's nest
235 ml milk
rock sugar
2 cups water

做法 Method

1. 燕窩用水浸約 3 小時,瀝乾水分,備用。

2. 煮滾水 2 杯,加入冰糖煮至溶。

3. 將木瓜頂部切去,刮掉木瓜籽,放入燕窩及糖水（七成滿）,蓋上木瓜頂部,放入燉盅內,以紗紙蓋好,隔水燉 45 分鐘,享用時加入鮮奶拌勻。

1. Soak bird's nest in water for 3 hours. Drain and set aside.

2. Boil 2 cups of water in a pot. Add rock sugar and cook till it dissolves.

3. Cut off the tops of the papayas and save them for later use. Remove the seeds. Fill the papayas with bird's nest and syrup up to 70% full. Cover with the tops. Put them into a double-steaming pot. Cover with mulberry paper. Steam for 45 minutes. Stir in milk before serving.

若經濟實惠的話，以雪耳代替燕窩，食療功效
相同。

For a frugal variation, use white fungus instead of
bird's nest. It tastes equally great and has similar
medicinal value.

鴛鴦芝麻千層糕

Black sesame and coconut milk layered cake

Video

材料 Ingredients （可製成8吋糕盆）
(makes one 8-inch cake)

泰國薯粉 2 杯
粘米粉 3 安士（量杯計）
糖 6 安士
鹽 1/8 茶匙
椰漿 400 毫升
芝麻糊粉 2 包（約 60 克）
水半杯
2 cups Thai tapioca starch
85 ml long-grain rice flour (measured in cup)
170 g sugar
1/8 tsp salt
400 ml coconut milk
60 g ground black sesames
1/2 cup water

❧ 小煮意 ❧

- 加入芝麻糊粉後或呈粒狀，必須篩勻或過濾；芝麻糊粉可先煮成糊狀，較容易與粘米漿拌和。

- 先將糕盆蒸熱，粉漿蒸熟後不易黏着糕盆。

- 每層不同顏色的粉漿要平均倒入，令每層厚薄相同，加蓋，各蒸 4 分鐘即可。

- If the mixture turns lumpy after you add the ground black sesames, you must strain the mixture through a sieve. To make the ground black sesames blend better with the batter, you may cook the ground black sesames with some water into a thick paste first.

- To turn the cake out more easily from the tray, you may steam the tray till hot before you pour in the first layer of batter.

- For better presentation, make sure each layer of batter is of similar thickness. After pour a new layer, cover the lid and steam for 4 minutes.

做法 Method

1. 粘米粉及薯粉篩勻。

2. 煮滾水半杯，加入糖及鹽攪拌，待涼，將乾粉分數次加入拌勻，下椰漿拌至幼滑，平均分成兩份。

3. 芝麻糊粉分數次加入其中一份粉漿內，拌勻備用。

4. 糕盆塗抹油，先倒入一層芝麻糊粉漿，蒸 4 分鐘後，再倒入粘米粉漿，再蒸 4 分鐘，相間地倒入兩種顏色粉漿，直至倒入最後一層粉漿後，蒸 20 分鐘即成。

1. Sieve rice flour and tapioca starch together.

2. Boil 1/2 cup of water in a pot. Add sugar and salt. Stir well. Let cool. Add a little dry ingredients from step 1 at a time and stir well after each addition. Add coconut milk and stir until well combined. Divide into 2 equal portions.

3. Stir in a little ground black sesames at a time into one portion of the batter. Mix well.

4. Grease a steaming tray. Pour in one layer of black sesame batter. Steam for 4 minutes. Then top with a layer of coconut milk batter. Steam for 4 minutes. Repeat this step to build alternate layers of black and white batter until you finish all ingredients. Then steam the whole cake for 20 more minutes. Serve.

Ginger coconut glutinous rice cake
with black sugar

黑蔗糖薑汁椰汁年糕

材料 Ingredients

糯米粉 4 杯
黑蔗糖 12 兩
椰汁 12 安士
薑汁 6 安士
花生油 2 湯匙
清水 3 安士

4 cups glutinous rice flour
450 g muscovado sugar
350 ml coconut milk
175 ml ginger juice
2 tbsp peanut oil
85 ml water

做法 Method

1. 燒熱水，放入黑蔗糖用慢火煮溶，待至和暖。

2. 糯米粉放入大盆內，逐少加入黑蔗糖水慢慢搓至粉糰滑身即可（約半小時），隨後下椰汁搓揉，再加入薑汁搓至滑身，下油搓半小時，最後加入餘下之黑蔗糖水拌勻。

3. 糕盆內塗上油，倒入粉糊，用大火隔水蒸約 2 至 3 小時即可。

1. Boil water and add muscovado. Cook until it dissolves. Leave it to cool until lukewarm.

2. Put glutinous rice flour into a mixing bowl. Slowly stir in warm muscovado syrup and knead into sticky dough until smooth (about 30 minutes). Add coconut milk and keep kneading. Add ginger juice and knead until smooth. Add oil and knead for 30 minutes. Put in the remaining muscovado syrup and mix well.

3. Grease a baking tray. Pour in the batter. Steam over high heat for 2 to 3 hours. Serve.

⤳ 小煮意 ⤳

- 建議黑蔗糖水逐少加入，花點時間搓成軟滑粉糰，能使粉糰均勻，年糕不會太軟太黏。

- 建議用慢火煮黑蔗糖，否則容易焦燶。

- I prefer slowly kneading in the muscovado syrup. That would ensure the dough is well mixed and the glutinous rice cake won't be too soft or too sticky.

- Cook the muscovado sugar over low heat, or it will burn easily.

良鄉栗子糊

Chestnut sweet soup

〜 小煮意 〜

栗子用滾水煮片刻，用毛巾包裹輕擦可去掉外衣。

To peel the chestnuts easily, blanch them in boiling water briefly. Then drain and wrap them in a towel. The skin can be rubbed off with a towel easily.

材料 Ingredients

栗子 12 兩
450 g chestnuts

調味料 Seasoning

冰糖 6 兩
栗粉 3 湯匙
水 4 湯匙
225 g rock sugar
3 tbsp cornstarch
4 tbsp water

做法 Method

1. 栗子去殼、去衣，洗淨，用水煲熟，以刀背壓成栗子茸。
2. 燒熱水 6 碗，放入冰糖煮至溶化，加入栗子茸攪勻，用慢火煮滾。
3. 栗粉及水調勻，徐徐拌入栗子茸內（邊加入邊攪拌），煮滾後即可享用。

1. Shell and peel the chestnuts. Rinse well. Boil them in water until cooked through. Crush the chestnuts with the back of a knife and mash very finely.

2. Boil 6 bowls of water. Put in rock sugar and cook until it dissolves. Stir in the mashed chestnuts. Bring to the boil over low heat.

3. Mix cornstarch with water. Slowly stir into the chestnut sweet soup. Keep stirring until it boils. Serve.

荔枝牛奶布甸

Lychee panna cotta

材料 Ingredients

荔枝肉 10 粒
魚膠粉 1 1/2 湯匙
凍開水 3 湯匙
鮮奶 800ml（26 安士）
糖 100 克（4 安士）
淡忌廉 300ml
荔枝香油半茶匙

10 lychees (shelled and pitted)
1 1/2 tbsp gelatine powder
3 tbsp cold drinking water
800 ml milk
100 g sugar
300 ml whipping cream
1/2 tsp lychee essence

做法 Method

1. 魚膠粉及凍開水拌勻，座於熱水待溶，備用。

2. 鮮奶及糖用小火煮熱（切記不要煮滾），待糖溶化後，熄火，倒入魚膠粉水拌勻。

3. 灑入荔枝香油、淡忌廉攪拌，最後加入荔枝肉，冷藏至凝固享用。

1. Mix gelatine powder with cold drinking water. Heat them up over a pot of simmering water until gelatine dissolves.

2. In a pot, put milk and sugar over low heat. Cook until sugar dissolves, but do not boil the milk. Turn off the heat. Stir in gelatine solution from step 1.

3. Add lychee essence and whipping cream. Mix well. Put in the lychees. Refrigerate till set. Serve.

195

蒸夾心雞蛋糕

Steamed cake with lotus seed paste filling

材料 Ingredients

雞蛋 5 個
糖 4 安士（量杯計）
麵粉 9 安士（量杯計）
泡打粉 1 茶匙
鹹蛋 2 個
蓮蓉或豆沙 4 兩
粟米油或牛油 2 湯匙
芫茜葉少許

5 eggs
115 ml sugar (measured in cup)
265 ml flour (measured in cup)
1 tsp baking powder
2 salted eggs
150 g lotus seed paste (or red bean paste)
2 tbsp corn oil or melted butter
coriander leaves

做法 Method

1. 煲滾水，下鹹蛋煲熟，取蛋黃切粒。

2. 蓮蓉壓成薄片，備用。

3. 雞蛋及糖放入大碗內，用打蛋器拌打至鬆軟，加入油略拌，分數次放入已篩勻之麵粉及泡打粉，略攪拌。

4. 糕盆放上牛油紙，倒入半份粉漿用中大火隔水蒸約 6 分鐘，成蛋糕底層。

5. 在蛋糕底層放上蓮蓉片，再倒入餘下之粉漿，加上蛋黃粒，繼續蒸約 8 分鐘，以芫茜葉裝飾即成。

1. Boil water in a pot. Put in the salted egg and cook it through. Shell and use only the egg yolk. Dice it finely.

2. Roll the lotus seed paste into a thin sheet slightly smaller than your cake tin. Set aside.

3. In a mixing bowl, put in eggs and sugar. Beat with electric mixer until fluffy. Add oil or butter. Stir a few times. Sieve in flour and baking powder a little at a time. Fold gently. Do not over-stir.

4. Line the cake tin with baking paper. Pour in half of the batter. Bake over medium-high heat for 6 minutes.

5. Put the lotus seed paste over the half-set cake base. Top with the remaining batter. Put diced salted egg yolk over. Steam for 8 more minutes. Garnish with coriander leaves. Serve.

✤ 小煮意 ✤

必須將蓮蓉平放於蛋糕底層，而且動作迅速，以免蛋糕漿流瀉。

The lotus seed paste should be placed flat on the cake base. Make sure
you do it quickly so that the batter won't overflow.

蜂蜜綠茶凍糕 Honey green tea jelly

❧ 小煮意 ❧

可不加入綠茶醬，效果相差不遠。綠茶醬於售賣蛋糕材料店有售。

You may omit green tea paste if you can't find it. The jelly won't taste too different without it. You can find green tea paste in baking supply stores.

材料 Ingredients

綠茶粉 1 茶匙
綠茶醬半茶匙
魚膠粉 5 湯匙
蜜糖 4 湯匙
糖 2 湯匙
水 2 杯
1 tsp green tea powder
1/2 tsp green tea paste
5 tbsp gelatine powder
4 tbsp honey
2 tbsp sugar
2 cups water

做法 Method

1. 煮滾清水 2 杯，加入魚膠粉、糖及綠茶粉用慢火煮溶，熄火，下蜜糖及綠茶醬拌勻。

2. 倒入糕盆內，冷藏至凝固，切成小件，伴生果享用。

1. Boil 2 cups of water and put in gelatine, sugar and green tea powder. Cook over low heat until all ingredients dissolve. Turn off the heat. Add honey and green tea paste.

2. Pour into a cake tine. Refrigerate until set. Slice and serve with fruits.

薑汁撞奶

Ginger milk custard

材料 Ingredients

蒙牛奶 250ml
老薑 1 大塊
糖 1 湯匙
250 ml Mengniu milk
1 large chunk old ginger
1 tbsp sugar

做法 Method

1. 老薑去皮，洗淨，磨成薑汁約 1 1/2 湯匙，盛
   於碗內。

2. 蒙牛奶放入煲內，略滾，來回多次倒入量杯，
   然後從高處倒入薑汁的碗內，加蓋焗 5 至 10
   分鐘，待凝固即可。

1. Peel the ginger and rinse well. Grate it and
   squeeze out the juice. You'd need 1 1/2 tbsp
   of ginger juice. Save in a bowl.

2. Pour milk and Mengniu milk into a pot.
   Bring to a gentle simmer. Pour the milk in the
   measuring cup a few times. Then pour the
   milk into the bowl with ginger juice. Cover
   the bowl and leave it for 5 to 10 minutes until
   set. Serve.

❧ 小煮意 ❧

磨好的薑汁應盡快以鮮奶撞
入，較易凝固。

The custard tends to set more
easily if the ginger juice is
freshly squeezed right before
you pour the hot milk in.

椰汁紅豆糕

Coconut milk cake
with red bean filling

粉漿料 Ingredients

粘米粉 10 安士（量杯計）
水 10 安士
椰漿 12 安士
紅豆 3 兩

295 ml long-grain rice flour
(measured in cup)
295 ml water
350 ml coconut milk
113 g red beans

糕面料 Brown sugar topping

片糖 1 塊
粘米粉 3 安士（量杯計）
水 8 安士

1 raw cane sugar slab
85 ml long-grain rice flour
(measured in cup)
235 ml water

煮紅豆料 Syrup

水 8 安士
冰糖 4 兩
油 1 湯匙

235 ml water
150 g rock sugar
1 tbsp oil

❧ 小煮意 ❧

- 購買天津紅豆，容易煲至軟腍起沙。

- 可用其他豆類如眉豆、綠豆、紅腰豆或鷹咀豆代替紅豆。

- 若想簡化步驟，可刪掉片糖糕面的製作。

- I prefer Tianjin red beans for this recipe. They cook more quickly and tend to be starchy.

- You may also use other beans for this recipe, such as black-eyed beans, mung beans, red kidney beans or chick peas.

- To simplify the steps, you may skip the brown sugar topping layer.

糕面做法 Method for brown sugar topping

1. 水 4 安士與粘米粉 3 安士調勻。
2. 煲滾 4 安士水，放入片糖煲至溶，倒入粉漿內拌勻，備用。

1. Mix half of the water with long-grain rice flour.

2. Boil the sugar slab in the remaining water in a pot. Pour in the slurry from step 1. Mix well and set aside.

做法 Method for coconut milk red bean cake

1. 紅豆洗淨，用水浸 3 小時，瀝乾水分。燒熱水 4 杯，加入紅豆用中火煮沸，轉慢火煲至紅豆開花（約 30 分鐘），隔水備用。
2. 燒熱水 8 安士，加入冰糖煮至溶，倒入油及紅豆略煮，備用。
3. 粘米粉倒入大碗內，調入水 10 安士開成粉漿，轉入煲內，加入椰漿以慢火煮成半生熟粉漿，下紅豆拌勻，倒入已抹油的糕盆內，用大火蒸半小時。
4. 倒入片糖漿於糕面上，再蒸半小時，取出待涼，切件享用。

1. Rinse the red beans and soak them in water for 3 hours. Drain. Boil 4 cups of water. Put in red beans and bring to the boil over medium heat. Turn to low heat and cook until red beans start to break up (for about 30 minutes). Drain.

2. Boil 235 ml of water in a pot. Add rock sugar and cook until it dissolves. Add oil and red beans from step 1. Cook briefly.

3. In a large mixing bowl, put in rice flour and 295 ml of water. Mix well and save in a pot. Add coconut milk and cook over low heat until half-set. Stir in red beans and mix well. Pour into a greased steaming tray. Steam over high heat for 30 minutes.

4. Pour the brown sugar topping over the steamed cake. Steam for 30 more minutes. Let cool and slice. Serve.

Steamed banana cake 蒸香蕉蛋糕

材料 Ingredients

麵粉 9 安士（量杯計）
發粉 1 1/2 茶匙
梳打粉半茶匙
牛油 4 安士
糖 6 安士
雞蛋 3 個（蛋黃及蛋白分開）
香蕉 2 隻
香蕉油 1 茶匙
265 ml flour (measured in cup)
1 1/2 tsp baking powder
1/2 tsp baking soda
112 g butter
165 g sugar
3 eggs (yolks and whites separated)
2 banana
1 tsp banana essence

ᘓᕽ 小煮意 ᕽᘐ

- 建議選較成熟的香蕉，製成的蛋糕香蕉味濃郁。

- 可以不拌入香蕉油，悉隨尊便。

- 除了烤焗之外，麵糊也可隔水蒸 30 分鐘至熟，同樣美味。

- I prefer using ripe bananas for this recipe so that the cake has rich fruity aroma.

- Banana essence is optional. You may skip it if you want.

- Apart from baking the cake, you may also steam it for 30 minutes. It tastes equally great.

做法 Method

1. 麵粉、發粉及梳打粉篩勻,備用。
2. 香蕉壓成泥糊狀,備用。
3. 牛油及糖打至奶白色,蛋黃分兩次加入,打勻。
4. 將香蕉茸及粉料分三次交錯地拌入牛油蛋黃料內,拌勻,再加入香蕉油攪勻。
5. 將蛋白打起,放入步驟4之香蕉麵糊輕手攪勻,倒入已墊紙及塗油的糕盆內。
6. 預熱焗爐180℃,放入糕盆焗25分鐘,以竹籤測試沒有黏着粉漿即成。

1. Sieve flour, baking powder and baking soda together.

2. Skin and mash the bananas.

3. Beat butter and sugar until pale. Put in 1 1/2 egg yolk at one time. Beat well after each addition.

4. Put in 1/3 of the mashed banana into the egg yolk mixture. Mix well. Put in 1/3 of the dry ingredients from step 1. Mix well. Repeat until all mashed bananas and dry ingredients are added and mixed in. Add banana essence. Mix again.

5. Beat egg whites until stiff. Fold it into the banana batter from step 4. Pour into a baking tray that is greased and lined with baking paper.

6. Preheat an oven to 180℃. Bake the cake for 25 minutes. Check doneness by inserting a toothpick at the centre of the cake. The cake is done if the toothpick comes out clean without any streak of uncooked batter. Serve.

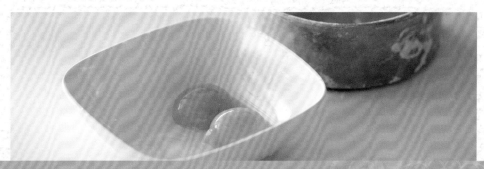

杞子桂花糕

Osmanthus jelly with Qi Zi

材料 Ingredients

蒟蒻凍粉 4 1/2 湯匙
糖 6 湯匙
杞子 20 粒
糖桂花 1 至 2 湯匙
清水 18 安士
4 1/2 tbsp konjac powder
6 tbsp sugar
20 Qi Zi
1 to 2 tbsp candied osmanthus
530 ml water

做法 Method

1. 蒟蒻凍粉及糖拌勻，注入清水開火不停攪拌，煮至糖溶化，加入杞子及糖桂花拌勻。

2. 傾入容器內，放入雪櫃冷凍至凝固即可。

1. Mix konjac powder with sugar in a pot. Add water and turn on the heat. Keep stirring until sugar dissolves. Stir in Qi Zi and candied osmanthus.

2. Pour into jelly cups or container of your choice. Refrigerate until set. Serve.

∽ 小煮意 ∽

若以乾桂花代替糖桂花，味道及口感略遜。

You may use dried osmanthus instead of candies osmanthus. But the jelly won't taste as great.

Coconut cream mousse

材料 Ingredients

蛋白 5 個	5 egg whites
椰汁 2 杯	2 cups coconut milk
鮮奶 1 盒 (235ml)	235 ml milk
魚膠粉 6 湯匙 (約 50 克)	6 tbsp (50 g) gelatine powder
糖 1 1/4 杯	1 1/4 cup sugar
水 2 杯	2 cups water

做法 Method

1. 清水 2 杯加入魚膠粉和糖煮溶。

2. 慢慢拌入椰汁及鮮奶攪勻，熄火，待涼。

3. 加入已打至企身之蛋白，拌勻後傾入容器內，放入雪櫃冷藏 2 至 3 小時至凝固，即可享用。

1. In a pot, add gelatine powder and sugar to 2 cups of water. Cook till sugar dissolves.

2. Slowly stir in coconut milk and milk. Turn off the heat and let cool.

3. Beat the egg whites until stiff. Fold the egg white into the coconut milk mixture from step 2 until well incorporated. Put into the container. Refrigerate for 2 to 3 hours until set. Serve.

❦ 小煮意 ❧

蛋白會慢慢升至上層，口感像棉花糖，軟綿可口；但緊記打蛋白的容器必須清潔及乾淨。

Beaten egg whites will rise to the top layer after stirring well. The egg whites give this dessert a light fluffy texture like marshmallow. Just make sure the utensils you use to beat the egg whites must be completely grease free. Otherwise, the egg whites won't stand.

Coconut milk noodles in mango puree

鮮芒果椰汁河粉

材料 Ingredients
蒟蒻凍粉 8 湯匙
糖 2 安士
椰漿 8 安士
鹽 1/8 茶匙
芒果粒 8 安士
芒果汁 8 安士
椰茸適量
芝麻適量
清水 10 安士
8 tbsp konjac powder
55 g sugar
235 ml coconut milk
1/8 tsp salt
225 g diced mango
235 ml mango puree
dried grated coconut
sesames
295 ml water

糖漿 Syrup
清水 2 安士
糖 2 安士
* 煮溶
60 ml water
55 g sugar
* cooked until sugar dissolves

❧ 小煮意 ❧

- 椰漿加入鹽調味，能夠帶出椰子的香味。

- 切椰汁河粉時，盡量切成薄條，仿似河粉般的質感。

- Adding salt to the coconut milk noodle batter helps bring out the coconut aroma.

- When you cut the coconut milk noodles, try to cut as thinly as you can, so that they resemble ribbon rice noodles.

做法 Method

1. 蒟蒻凍粉及糖拌勻，注入清水內，加入椰漿及鹽攪勻，煮至糖溶化。

2. 傾入長方型盤內，冷藏至凝固，切成幼條狀。

3. 將椰汁河粉放於碟內，澆上糖漿、芒果汁及芒果粒，最後撒上椰茸及芝麻即可品嘗。

1. In a pot, mix konjac powder with sugar. Add water and stir well. Then add coconut milk and salt. Turn on the heat. Cook while stirring until sugar dissolves.

3. Pour the mixture into a rectangular tray. Refrigerate until set. Cut into thin strips.

4. Put coconut milk noodles on a serving plate. Drizzle with syrup, mango puree and diced mango. Sprinkle with dried grated coconut and sesame on top. Serve.

Sticky rice pancakes with
red bean filling

豆沙煎軟糍

材料 Ingredients

糯米粉 16 安士（量杯計）
澄麵 4 湯匙
豬油 2 湯匙
糖 2 安士
水 6 安士
滾水 2 安士
白芝麻 2 湯匙

470 ml glutinous rice flour
(measured in cup)
4 tbsp wheat starch
2 tbsp lard
55 g sugar
175 ml water
60 ml boiling water
2 tbsp white sesames

餡料 Filling

豆沙 1 斤
鹹蛋黃 3 個

600 g red bean paste
3 salted egg yolks

小煮意

澄麵煮熟後，熄火，加蓋待片刻，澄麵糰才變得透明。

After cooking the wheat starch, you have to turn off the heat and leave the dough in the pot with the lid covered for a while. This step makes the wheat starch look translucent.

做法 Method

1. 糯米粉篩勻於大碗內，逐少加入水搓成軟麵欄。
2. 澄麵篩勻，倒入滾水內拌勻煮熟，加蓋待片刻。
3. 將澄麵欄加入糯米粉欄內一同搓勻，加入豬油及糖再搓至軟滑。
4. 鹹蛋黃蒸熟，每個切成8小粒，用豆沙包着鹹蛋黃如波子大小粒狀。
5. 將糯米麵欄搓成長條，切成小粒，以手搓成窩形，放入豆沙鹹蛋黃，收口後壓扁，餅面灑上白芝麻。
6. 燒熱平底鑊，下少許油，放入小圓餅煎至兩面金黃色即成。

1. Sieve the glutinous rice into a big bowl. Slowly add water and knead into smooth dough.

2. Sieve the wheat starch. Add to 60 ml of water that has recently been boiled in a pot and mix well. Cover the lid and leave it briefly.

3. Put the wheat starch dough into the glutinous rice dough. Knead to mix well. Add lard and sugar. Knead until smooth.

4. Steam the salted egg yolks until done. Cut into 8 small pieces. Wrap each piece of salted egg yolk in a small piece of red bean paste. It should be the size of a marble.

5. Roll the dough into a long cylinder. Cut into small pieces. Roll each piece out into a small bowl. Put in the red bean and salted egg yolk filling. Seal the seam and press flat into a patty. Sprinkle with white sesames.

6. Heat a pan and add a little oil. Fry the pancakes until both sides golden. Serve.

糖不甩 Glutinous rice balls with peanut sesame topping

材料 Ingredients

糯米粉 3 杯
暖水 9 安士
片糖 4 片
清水 1 杯
3 cups glutinous rice flour
265 ml warm water
4 raw cane sugar slabs
1 cup water

配料 Toppings

炒香花生茸 2 兩
炒香白芝麻 2 兩
椰茸 2 兩
糖 2 兩
75 g ground toasted peanuts
75 g toasted white sesames
75 g dried grated coconut
75 g sugar

做法 Method

1. 糯米粉篩勻，放於盆內，中間挖一個洞，逐少加入暖水，搓成粉糰。
2. 燒滾清水 1 杯，下片糖煲至溶化。
3. 將麵糰搓成小粒狀，逐粒放入片糖水內煮至浮起，隔起，撒上配料享用。

1. Sieve the glutinous rice flour into a tray or big bowl. Make a well at the centre. Add warm water into the well a little at a time. Knead after each addition into smooth dough.

2. Boil 1 cup of water in a pot. Put in the raw cane sugar slabs. Cook until they dissolve.

3. Divide the dough from step 1 and roll into small balls. Put them into the boiling syrup from step 2. Cook until they float. Remove with a strainer ladle and transfer onto serving plate. Sprinkle with the toppings and serve.

≈ 小煮意 ≈

煮片糖宜用小火慢煮，否則糖水容易變焦。

Always cook raw cane sugar syrup over low heat. Or else it may burn.

笑口棗　Deep-fried crunchy sesame balls

材料 Ingredients

麵粉 34 安士（量杯計）	4 cups flour
梳打食粉 1 茶匙	1 tsp baking soda
發粉 1 茶匙	1 tsp baking powder
豬油 2 湯匙	2 tbsp lard
食用鹼水半茶匙	1/2 tsp food-grade lye
白芝麻 1 杯	1 cup white sesames
糖 1 1/2 杯	1 1/2 cups sugar
水 7 安士	200 ml water

做法 Method

1. 煮滾水，下糖煮溶，待涼備用。

2. 麵粉篩勻，放於桌面，中間開成一穴，放入豬油、發粉及梳打粉慢慢搓勻，下糖水及鹼水略拌，用粉刀將麵粉拌入糖水內，略揉至軟滑，分成三份。

3. 取一份粉糰用手壓平，疊上第二份粉糰再略揉，最後放上第三份粉糰，搓成長條，切粒，沾少許水，灑上白芝麻，用乾布蓋上發酵半小時。

4. 燒熱油，調至小火，放入小粉糰，見浮起時調中火炸至金黃色即成。

1. Boil water. Put in sugar and cook until it dissolves. Let cool.

2. Sieve flour onto a counter. Make a well at the centre. Put in lard, baking powder and baking soda. Slowly knead well. Add sugar syrup from step 1 and food-grade lye. Stir slightly. Push the flour into the well with a dough cutter. Knead into soft dough. Divide into 3 equal portions.

3. Press one portion flat with your palm. Top with the second portion and knead gently. Put the third portion on top at last. Roll into a long cylinder. Cut into pieces. Dip each piece in water and roll it in white sesames. Cover with a dry towel and let it rise for 30 minutes.

4. Heat oil till hot. Turn to low heat. Put in the dough balls. Deep fry until they float. Turn up the heat to medium and fry until golden. Drain and serve.

❧ 小煮意 ❧

- 炸笑口棗時，不時撥動滾油，令笑口棗的色澤均勻。

- 食用鹼水於雜貨店有售。

- When you deep-fry the sesame balls, make sure you roll them in the oil from time to time so that they pick up the golden colour evenly.

- Food-grade lye is available from grocery stores.

Sweet potato Jian Dui
(Deep-fried glutinous rice sesame balls)

炸番薯煎堆

材料 Ingredients

番薯 1 斤
糯米粉 1 斤
片糖 4 兩
砂糖 1 湯匙（搓皮用）
白芝麻少許
600 g sweet potatoes
600 g glutinous rice flour
150 g raw cane sugar slabs
1 tbsp sugar (for kneading)
white sesames

餡料 Filling

豆沙或蓮蓉 1 斤
600 g red bean paste (or
lotus seed paste)

❧ 小煮意 ❧

- 選黃肉番薯，令煎堆香甜美味。

- 粉糰加入砂糖搓勻，令外皮香脆。

- 要待番薯片糖水和暖才加入糯米粉內，否
  則容易將糯米粉弄熟。

- 愛吃鹹味的話，可換成鹹味餡料：臘腸 2 條、
  臘肉半條、蝦米 1 兩、菜甫 4 條、冬菇 4
  朵（浸軟去蒂、蒸熟）、馬蹄 8 粒、葱粒
  4 湯匙，全部切粒，炒勻成餡料。

- It's best that you use yellow sweet potato
  for this recipe because of its sweetness and
  starchiness.

- Adding 1 tbsp of sugar to the dough helps
  make it crispier.

- Wait till the cane sugar syrup is warm before
  adding it to the flour. Otherwise, you'd cook
  the flour right away and it turns lumpy.

- If you prefer savoury glutinous rice balls,
  try to use the followings as the filling: 2
  Chinese preserved pork sausages, 1/2 strip
  Chinese preserved pork belly, 38 g dried
  shrimps, 4 pieces dried radish, 4 dried shiitake
  mushrooms (soaked in water till soft, stems
  removed and steamed till done), 8 water
  chestnuts, 4 tbsp diced spring onion. Just chop
  all ingredients up and stir-fry.

做法 Method

1. 番薯去皮、切塊,加入水3碗煲熟,下片糖煮溶,將番薯放入已篩勻的糯米粉內搓勻,慢慢加入片糖水,搓至不黏手,最後下砂糖搓至軟滑。

2. 將番薯麵糰分成小粒,略壓平,包入餡料,搓圓,滾上芝麻。

3. 燒熱油,放入番薯麵糰,用杓子輕輕轉動,改用小火待麵糰炸至金黃色,見浮面調至大火即可盛起,瀝乾油分,趁熱享用。

1. Peel the sweet potatoes. Cut into pieces. Boil 3 bowls of water and cook them until done. Add raw cane sugar slabs and cook until they dissolves. Sieve rice flour into a mixing bowl. Put in the sweet potatoes. Knead well. Slowly stir in the sugar syrup. Knead until the dough no longer sticks to your hand. Add 1 tbsp of sugar and knead until smooth.

2. Divide the dough into small pieces and press each piece flat. Wrap some filling in it. Seal the seam and roll it round. Then roll it in sesames to coat well.

3. Heat oil in a wok and deep fry the sweet potato balls over low heat until golden. Keep turning them with a spatula throughout the process. Turn to low heat and deep-fried until golden. They are cooked through when they float on the oil. Turn to high heat for a while. Drain and serve hot.

擂沙煎堆

少變化，多口味

材料 Ingredients

糯米粉 4 杯
發粉 2 茶匙
片糖 3 片
白芝麻 1 杯
砂糖 1 湯匙（搓皮用）
水 12 至 14 安士
4 cups glutinous rice flour
2 tsp baking powder
3 raw cane sugar slabs
1 cup white sesames
1 tbsp sugar (for kneading)
350-410 ml water

餡料 Filling

黑麻茸 1 斤
600 g black sesame paste

Jian Dui with black sesame filling

做法 Method

1. 燒滾水，加入片糖煮溶，待至暖和。

2. 糯米粉篩勻於桌面，在中間開成一穴，加入發粉，逐少加入糖水及糖1湯匙搓勻，搓至粉糰幼滑，用濕布蓋着待半小時。

3. 將麵糰搓成長條狀，切成小糯米糰，略按扁，包入黑麻茸，搓成圓球狀，滾上芝麻。

4. 油略燒熱，放入小麵糰用小火炸至呈金黃色，見表面漲大時，改大火再炸片刻，瀝乾油分即可。

1. Boil water and put in the raw cane sugar slabs. Cook until they dissolves. Let the syrup cool down.

2. Sieve glutinous rice flour on a counter. Make a well at the centre. Put in baking powder. Add the warm syrup from step 1 and 1 tbsp of sugar. Knead until smooth. Cover with a damp towel for 30 minutes.

3. Roll the dough into a long cylinder. Cut into small pieces. Press each piece flat slightly. Wrap in some black sesame paste. Seal the seam and roll into a ball. Roll the dough ball in white sesames to cover evenly.

4. Heat oil in a wok or fryer. Put in the dough balls and deep fry over low heat until golden. When they start to puff up, turn up the heat and fry for a while. Drain and serve.

南瓜煎堆

少變化，多口味

材料 Ingredients
南瓜 1 斤
糯米粉 1 斤
片糖 4 兩
砂糖 1 湯匙（搓皮用）
白芝麻 1 杯
水 3 杯
600 g pumpkin
600 g glutinous rice flour
150 g raw cane sugar slabs
1 tbsp sugar (for kneading)
1 cup white sesames
3 cups water

餡料 Filling
豆沙或蓮蓉 1 斤
600 g red bean paste
(or lotus seed paste)

Pumpkin Jian Dui

做法 Method

1. 南瓜去皮、切塊，加入水2杯煲熱，下片糖煮溶，待至和暖。

2. 將南瓜放入已篩勻的糯米粉內搓勻，慢慢加入片糖水及糖，搓至不黏手及軟滑。

3. 南瓜麵糰搓成長條，切成如湯丸般小粒，略壓扁，包入餡料，搓成棗形，灑上少許水，滾上芝麻。

4. 燒熱油，放入南瓜球，用杓子輕輕轉動，改用小火待麵糰炸至金黃色，轉大火炸一會，瀝乾油分，上碟即成。

1. Peel pumpkin and cut into piece. Boil 2 cups of water and cook the pumpkin through. Add raw cane sugar slabs and cook until they dissolve. Leave to cool down.

2. Sieve glutinous rice flour into a mixing bowl. Put in the pumpkin. Knead to mix well. Then add the warm syrup from step 1 and 1 tbsp of sugar. Knead until the dough no longer sticks to your hand and is smooth.

3. Roll the dough into a long cylinder. Cut into small pieces. Press each piece flat slightly. Wrap in some filling and roll into an elongated oval with pointy ends. Sprinkle with water and roll them in sesames to coat well.

4. Heat oil in a wok or fryer. Deep fry the pumpkin balls over low heat until golden. Turn them once in a while with a spatula. Turn the heat up to high at last and fry for a while longer. Drain and serve.

甜薄𧋘 Chinese crepe with sweet filling

材料 Ingredients
糯米粉 4 安士 (量杯計)
泰國薯粉 4 安士 (量杯計)
雞蛋 4 個
鮮奶 6 安士
清水 6 至 8 安士
糖 3 湯匙
115 ml glutinous rice flour (measured in cup)
115 ml Thai tapioca starch (measured in cup)
4 eggs
175 ml milk
175 to 235 ml water
3 tbsp sugar

餡料 Filling
花生茸、芝麻、
椰茸及糖各適量
ground peanuts
sesames
dried grated coconut
sugar

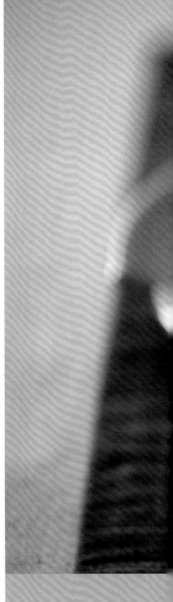

❧ 小煮意 ❧

薄𧋘煎至金黃色後，必須待涼才包入餡料，否
則糖會溶掉。

Make sure you wait till the crepe is cool before
you wrap the filling in. Otherwise, the sugar will
melt and the filling becomes sticky.

做法 Method

1. 將糯米粉、薯粉及糖放入盆內，加入雞蛋、鮮奶及清水攪拌至漿狀。

2. 燒熱平底鑊，下油 4 湯匙，傾出油分，舀入適量粉漿，煎至兩面金黃色，撒上餡料，捲起即可享用。

1. Mix glutinous rice flour, tapioca starch and sugar in a mixing bowl. Add eggs, milk and water. Stir into a runny thin paste.

2. Heat a pan and add 4 tbsp of oil. Swirl to coat evenly. Pour out the oil. Scoop in some batter with a ladle. Swirl the pan to spread evenly. Fry until both sides golden. Set aside to let cool. Sprinkle with fillings. Roll it up and serve.

豆沙角　Deep-fried red bean dumplings

材料 Ingredients
糯米粉 1 斤
粘米粉 4 湯匙
油 1 湯匙
水 2 杯
片糖 4 兩
600 g glutinous rice flour
4 tbsp long-grain rice flour
1 tbsp oil
2 cups water
150 g raw cane sugar slabs

餡料 Filling
豆沙 1 斤
600 g red bean paste

做法 Method

1. 燒滾水 2 杯，下片糖煮溶，備用。
2. 半份糖水趁熱倒入 6 兩糯米粉內，快手攪拌成糊狀。
3. 將以上的粉糊放入餘下的糖水內，煮片刻，熄火；再將粉糊放入 10 兩糯米粉及 4 湯匙粘米粉內拌勻，加入油 1 湯匙搓勻至軟滑。
4. 麵糰分成小份，以手揑薄邊沿，包入豆沙餡料，收口揑成角形，放入中火油鑊內炸至金黃色即可。

1. Boil 2 cups of water. Put in the raw cane sugar slabs. Cook until they dissolve.

2. Put 225 g of glutinous rice flour into a bowl. Pour in half of the hot syrup from step 1. Stir quickly into a sticky paste.

3. Then put the sticky paste from step 2 into the remaining syrup from step 1. Cook briefly and stir well. Turn off the heat. Put the resulting paste into the remaining glutinous rice flour. Add 4 tbsp of long-grain rice flour. Knead well. Add 1 tbsp of oil and knead into smooth dough.

4. Divide the dough into small pieces. Flatten and pinch the edge to thin it out. Wrap in some red bean paste filling. Seal the seam and shape it into dumpling with two pointy ends. Deep fry in hot oil over medium heat until golden. Serve.

小煮意

豆沙角的外皮用生、熟糯米粉拌勻，製成後帶煙韌口感。

In this recipe, part of the glutinous rice flour is pre-cooked. This gives the dumplings an extra-chewy texture.

Chocolate egg waffles

朱古力雞蛋仔

材料 Ingredients

粘米粉 5 湯匙
自發粉 4 安士（量杯計）
朱古力粒適量
雞蛋 3 個
糖 5 安士（140 克）
淡奶 2 湯匙
雲呢拿香油 1/4 茶匙
已溶牛油 2 湯匙
鹽 1/8 茶匙

5 tbsp long-grain rice flour
115 ml self-raising flour (measured in cup)
chocolate chips
3 eggs
140 g sugar
2 tbsp evaporated milk
1/4 tsp vanilla essence
2 tbsp melted butter
1/8 tsp salt

❦ 小煮意 ❦

配料可隨意添加，如果仁或黑芝麻
等，增加不同的口感。

You may add other ingredients such as
nuts or black sesames to the batter for
different textures.

做法 Method

1. 雞蛋及糖打至軟身，淡奶及雲呢拿香油分數次加入，備用。

2. 自發粉、粘米粉及鹽篩勻，慢慢加入雞蛋混合物內，以手拌勻，下牛油混和，待 15 分鐘。

3. 將雞蛋仔鐵模燒熱，抹上油，倒入蛋漿，灑上朱古力粒，蓋上模具，左右搖勻，用慢火每邊燒 3 分鐘，取出即成。

1. Beat eggs and sugar until stiff. Add evaporated milk and vanilla essence a little at a time. Set aside.

2. Sieve self-raising flour, rice flour and salt into a mixing bowl. Slowly stir in the egg mixture from step 1. Mix well. Add butter and mix again. Let it sit for 15 minutes.

3. Heat the egg waffle irons and grease them. Pour in the batter and sprinkle with chocolate chips. Cover the irons and jiggle it from side to side. Heat over low heat for 3 minutes on each side. Turn the waffle out and serve.

椰汁糯米糍

Coconut glutinous rice cake
with red bean filling

材料 Ingredients
椰汁 6 安士
糯米粉 8 安士（量杯計）
糖 5 安士
椰絲少許
水 3 安士
175 ml coconut milk
235 ml glutinous rice flour
(measured in cup)
140 g sugar
dried grated coconut
90 ml water

餡料 Filling
豆沙半斤
300 g red bean paste

小煮意

餡料可包入蓮蓉、粗粒花生醬、黑麻蓉或花生椰絲等。

You may also use lotus seed paste, crunchy peanut butter, black sesame paste or peanut and shredded coconut as filling.

做法 Method

1. 椰汁、糯米粉、糖及水拌勻，放於已塗油的碟上，隔水蒸半小時，待涼備用。

2. 豆沙分成 20 小粒，備用。

3. 將蒸熟的糯米糰分成 20 小粒，每粒包入豆沙，表面滾上椰絲，放在紙杯上即可。

1. In a mixing bowl, mix coconut milk, glutinous rice flour, sugar and water. Knead well. Put the dough on a greased plate. Steam for 30 minutes. Let cool.

2. Divide the red bean paste into 20 pieces.

3. Divide the steamed dough into 20 pieces. Press each piece flat and wrap in 1 piece of red bean filling. Seal the seam and roll them round. Roll them in dried grated coconut. Put into small paper cups and serve.

紅豆鉢仔糕
Put Chai Ko with red beans
(Steamed rice cake in bowls)

材料 Ingredients

紅豆 2 兩
粘米粉 20 安士（量杯計）
澄麵 3 湯匙
粟粉 3 湯匙
糖半斤（約 10 安士，或用片糖半斤）
水 5 1/2 杯

75 g red beans
590 ml long-grain rice flour (measured in cup)
3 tbsp wheat starch
3 tbsp cornstarch
300 g sugar (or raw cane sugar slabs)
5 1/2 cups water

做法 Method

1. 紅豆洗淨，燒滾水煲至腍，隔去水分。
2. 粘米粉、澄麵及粟粉篩勻，用水 1 1/2 杯調勻。
3. 燒滾水 4 杯，下糖煲至溶，倒入粉漿內攪勻，用密篩過濾一次。
4. 預先將小碗蒸熱，塗抹油，倒入粉漿，在表面放上紅豆，大火蒸 25 分鐘即可。

1. Rinse red beans and boil them in water until soft. Drain.

2. Sieve rice flour, wheat starch and cornstarch together into a mixing bowl. Add 1 1/2 cups of water. Mix well.

3. Boil 4 cups of water. Add sugar and cook until sugar dissolves. Pour the syrup into the batter from step 2. Mix well. Pass the batter through a sieve once.

4. Steam a few ceramic bowls until hot. Grease them. Pour in the batter. Put some red beans on top. Steam over high heat for 25 minutes until done. Serve.

～ 小煮意 ～

紅豆煲至腍身即可，毋須煲爛至開花，否則賣相不美。

Just boil the red beans until soft. They should still be in whole without breaking down. Otherwise, your Put Chai Ko won't look good.

酥炸角仔

Yau Gok
(Cantonese deep-fried oil dumplings)

材料 Ingredients
麵粉半斤
滾油 2 安士
凍油 2 安士
凍水 3.5 安士
300 g flour
60 ml boiling hot oil
60 ml cold oil
100 ml cold water

餡料 Filling
花生 2 兩
糖 1/4 杯
炒香白芝麻 1 兩
椰茸 1 兩
75 g peanuts
1/4 cup sugar
38 g toasted white sesames
38 g dried grated coconut

小煮意

- 先將滾油及凍油拌勻，再與凍水搓成皮，毋須加入豬油，也令炸出來的角仔皮脆可口。

- 傳統的做法是用豬油搓成外皮，香脆度十足，但今天為了健康着想，用生油代替效果也不錯。

- 搓麵糰時不要壓得太薄，否則酥皮的層數發不起來。

- Mixing hot oil and cold oil first, then add cold water separately with flour is my secret trick to crispy and flaky skin without the use of lard.

- Traditionally, the skin of the dumpling is made with lard which makes the skin flaky and aromatic. For those health-conscious, this lard-free version still tastes great.

- Do not roll the skin too thin. Otherwise, the layers of pastry skin cannot puff up.

做法 Method

1. 花生炒香後，去衣及壓碎，加入其他餡料拌勻，備用。

2. 將滾油及凍油拌勻；另備凍水一碗。

3. 麵粉篩勻放於桌面，中間開成一穴，逐少放入暖油及凍水，分數次搓成一小團麵糰，疊在一起，直至所有麵粉完成，搓成軟滑的麵糰。將麵糰切成三份，相疊後搓成麵糰（共兩次）。麵糰再切成四份，相疊後搓軟。

4. 麵糰搓成長條，分切成小粒，用擀木棒壓平，用模子壓出圓餅皮，包入餡料，揑成角形，放入170℃熱油炸至金黃色即可。

1. Toast the peanuts in a dry wok until fragrant. Peel and crush them. Mix in with other filling ingredients. Set aside.

2. In a mixing bowl, mix boiling hot oil and cold oil. Prepare a bowl of cold water.

3. Sieve flour onto a counter. Make a well at the centre. Slowly stir in the warm oil and cold water separately to make a small pieces. Place one by one and knead into soft dough. Cut the dough into three parts. Fold them and knead the dough again. Roll it out and repeat once. Finally, cut it into four and then fold it. Knead into soft dough.

4. Roll the dough into a cylinder. Cut into small pieces. Roll each piece out and cut out the skin with a cookie cutter. Wrap some filling in the skin. Crimp the seam. Deep fry in oil at 170°C until golden. Serve.

豆沙鍋餅

Deep-fried crispy pancake with red bean filling

∽ 小煮意 ∾

煎餅皮時，毋須反轉，煎一面即可。

When you fry the pancake for the first time, you don't need to flip it. Just fry it on one side.

材料 Ingredients
麵粉 4 安士（量杯計）
粘米粉 2 湯匙
雞蛋 2 個
豆沙 4 兩
凍水 9 安士

115 ml flour (measured in cup)
2 tbsp long-grain rice flour
2 eggs
150 g red bean paste
265 ml water

做法 Method

1. 麵粉及粘米粉置於大碗內，逐少雞蛋加入，拌勻，再逐少倒入凍水，不斷攪拌至滑糊狀。

2. 燒熱平底鍋，抹上少許油，待油燒熱後離火，倒入半杓粉漿，迅速轉動平底鍋，做成圓薄餅狀，用慢火煎熟。

3. 於餅皮的中間位置抹上一層豆沙，包成長方形，以粉糊黏口包好。

4. 燒熱油 5 湯匙，下鍋餅用中火半煎炸至金黃色，取出，切件享用。

1. Put flour and rice flour into a mixing bowl. Slowly pour in the eggs and mix well after each addition. Then slowly add cold water and keep stirring until smooth and lump-free.

2. Heat a pan and grease it with oil. Heat until oil is smoking hot and remove from heat. Pour in half a ladle of the batter from step 1. Quickly swirl the pan to coat evenly. Fry over low heat into a round pancake.

3. Spread a layer of red bean paste at the centre of the pancake. Fold the sides in toward the centre into a rectangular parcel. Seal the seams with a little batter.

4. Heat 5 tbsp of oil. Put the pancake in and fry over medium heat until golden. Cut into pieces and serve.

擂沙湯丸
Glutinous rice balls in ground peanut flour

材料 Ingredients
糯米粉 3 杯
糖 6 湯匙
豬油 1 湯匙
花生茸或黃豆茸半杯
水 9 安士
3 cups glutinous rice flour
6 tbsp sugar
1 tbsp lard
1/2 cup ground peanut flour (or
ground soybean flour)
265 ml water

餡料 Filling
黑麻茸半磅
225 g black sesame paste

做法 Method

1. 糯米粉篩勻放於大碗內，中間開成一穴，加入糖及豬油，水分數次加入搓溶，
   搓成軟滑麵糰。

2. 將麵糰搓成長條狀，切出小粒，用手壓成窩形，包入餡料，收口，搓成湯丸形狀。

3. 燒滾半鍋水，放入粉糰煮至浮起，取出滾上花生茸或黃豆茸即可。

1. Sieve the glutinous rice flour into a mixing bowl. Make a well at the centre.
   Add sugar and lard. Add water slowly to dissolve the sugar. Knead into soft
   smooth dough.

2. Roll the dough into a long cylinder. Cut into small pieces. Press each piece
   into a bowl shape. Wrap in filling. Seal the seam and roll into a ball.

3. Boil half a pot of water. Cook the dough balls until they float. Drain. Roll
   them in ground peanut flour or soybean flour. Serve.

小煮意

黑麻茸、花生茸或黃豆茸，在上海南貨店有售。

You can get black sesame paste, ground peanut or soybean flour from Shanghainese grocery stores.

蓮茸西米角

Sago dumplings with lotus seed paste filling

材料 Ingredients

西米半磅（約 10 安士）
糖 5 安士
粟粉 2 安士（量杯計）
泰國薯粉 2 安士（量杯計）
225 g sago
140 g sugar
60 ml cornstarch (measured in cup)
60 ml Thai tapioca starch (measured in cup)

餡料 Filling

蓮茸半斤
鹹蛋黃 3 個（切粒）
300 g lotus seed paste
3 salted egg yolks (diced)

❧ 小煮意 ❧

- 西米不可直接放於煲內煮，否則容易黏着底部。

- 煮西米時必須有耐性，邊煮邊不停攪拌，待西米呈半透明狀態。

- Never heat sago directly over the heat source. It tends to stick and burn easily.

- Be patient when you cook the sago. You must keep stirring until it turns translucent.

做法 Method

1. 西米用水浸 20 分鐘，盛起，隔水備用。

2. 西米、粟粉及糖放於攪拌盆內，鑊燒熱半鍋水，放入盆隔水以慢火不斷攪拌，直至全部拌勻及西米呈半熟，盛起。

3. 桌面灑上薯粉，放上西米糰搓揉至幼滑，分成 20 份小糰。用手壓平西米糰，放入蓮茸及鹹蛋黃，包捏成角形，放於已塗油的碟內，隔水蒸 10 分鐘即可。

1. Soak sago in water for 20 minutes. Drain.

2. Put sago, cornstarch and sugar into a big mixing bowl. Boil half a pot of water and heat the big bowl over the pot of simmering water. Mix the ingredients until well combined and the sago turns translucent. Set aside.

3. Flour your counter with tapioca starch. Put the sago mixture from step 2 on the counter and knead until smooth and not sticky. Divide into 20 equal portions. Press each piece of dough flat with your hand. Wrap some lotus seed paste and a piece of salted egg yolk in it. Seal the seam and shape it into a dumpling with pointy ends. Put on a greased plate. Steam for 10 minutes. Serve.

蒸南瓜餅

Steamed pumpkin cake

材料 Ingredients
南瓜 1 斤
糯米粉 3 杯
粘米粉 2 湯匙
糖 6 安士
糭葉 6 片
600 g pumpkin
3 cups glutinous rice flour
2 tbsp long-grain rice flour
170 g sugar
6 bamboo leaves

餡料 Filling
花生茸 1 杯
1 cup ground peanuts

做法 Method
1. 南瓜洗淨，去皮及去核，開邊，隔水蒸約 30 分鐘。
2. 糭葉洗淨，下滾水內煮片刻，剪成小片，掃上油備用。
3. 南瓜、糯米粉、粘米粉及糖拌勻，搓成軟粉糰，再分成數小粒。
4. 將小粉糰壓平，包入餡料後放入模內壓成型，排在糭葉上蒸 10 分鐘即成。

1. Rinse the pumpkin. Remove the skin and seeds. Cut in half. Steam for 30 minutes.

2. Rinse the bamboo leaves. Boil them in water for a while. Cut into small pieces. Brush oil on them.

3. In a mixing bowl, put in steamed pumpkin, glutinous rice flour, long-grain rice flour and sugar. Mix well. Then knead into soft dough. Divide into small pieces.

4. Press each piece of dough flat. Wrap in some filling. Seal the seam and press in the mould. Unmould and put on a piece of greased bamboo leaf. Steam for 10 minutes. Serve.

小煮意

將南瓜餅鋪在糭葉或蕉葉上蒸，可散發陣陣葉香味。

Steaming the pumpkin cake over a piece of bamboo or banana leaf adds an extra fragrance to the cake.

拔絲香蕉　Candied banana fritters

材料 Ingredients
香蕉2隻（大）
炒香白芝麻1湯匙
2 large bananas
1 tbsp toasted white
sesames

炸漿 Deep frying batter
麵粉8安士（量杯計）
泡打粉2茶匙
生粉3湯匙
水7安士
油1湯匙（後下）
235 ml flour (measured in cup)
2 tsp baking powder
3 tbsp caltrop starch
205 ml water
1 tbsp oil (added at last)

糖膠 Sugar coating
糖6安士
麥芽糖3湯匙
水4湯匙
170 g sugar
3 tbsp maltose
4 tbsp water

小煮意

煮糖膠時用慢火，以免焦燶；待糖及麥芽糖煮溶後，滴入冰水中，若見形成珠粒即可。

When you cook the sugar coating, make sure you heat it over low heat as sugar burns very easily. Cook till maltose and sugar melt. Test its consistency by dripping a drop of the syrup into ice water. It's done if it solidifies into a candy bead.

做法 Method

1. 炸漿材料拌勻待約半小時，最後下油輕拌。
2. 香蕉去皮，斜切成 2cm 厚件，沾上炸漿。
3. 燒熱半鑊油，放入香蕉炸漿，以大火炸至呈金黃色，隔去油分。
4. 將糖膠料放入鑊內，煮至呈金黃糖膠絲狀，放入炸好之香蕉拌勻，灑上芝麻。
5. 享用時，將香蕉放進冰水內，外層糖膠呈略硬，即成香脆的拔絲香蕉。

1. Mix the deep frying batter ingredients. Leave them for 30 minutes. Stir in the oil and mix well.

2. Peel the bananas. Slice into 2cm-thick pieces. Coat them in the batter.

3. Heat half a wok of oil. Deep fry the bananas over high heat until golden. Drain.

4. Put all sugar coating ingredients into the wok. Cook until it caramelizes into golden yellow and you can pull threads of sugar out. Put in the deep-fried bananas. Toss well. Sprinkle with sesames.

5. Transfer the bananas into ice water to crisp up the sugar coating. Put on a plate and serve.

Soufflé egg white balls
with red bean filling

高力豆沙

材料 Ingredients
麵粉 4 湯匙
蛋白 6 個
糖粉 4 湯匙
4 tbsp flour
6 egg whites
4 tbsp icing sugar

餡料 Filling
豆沙半斤
300 g red bean paste

蛋白打至鬆軟企身即可，別打得太久，否則欠鬆鬆的口感。

Do not over-whisk the egg whites. They should form soft peaks but not stiff. Otherwise, the soufflé balls will be too stiff instead of fluffy.

做法 Method

1. 將豆沙分成適量等份。
2. 蛋白拂打至鬆軟企身，加入已篩勻的麵粉，輕輕拌勻。
3. 將大湯匙放入熱油內，舀豆沙放入蛋白內滾圓，再放入熱油內，以中慢火炸至金黃色，上碟，灑上糖粉伴吃。

1. Divide the red bean paste into pieces roughly the same size.

2. Whisk the egg whites until soft peaks form. Sieve in flour and fold in gently.

3. Put a tablespoon into a pot of hot oil. Scoop out a tbsp of egg white. Put a piece of red bean paste at the centre. Roll it round and unload it into the hot oil. Fry over medium-low heat until golden. Save on a plate and sprinkle with icing sugar. Serve.

葉胡影儀女士
Mrs. Yip Woo Ying Yee

被學生稱為「鄰家的媽媽」的葉太,心靈手巧、廚藝精湛、熱愛烹飪,尤其炮製家常小菜更是拿手絕活。

廚藝顯真情

23 年以來,葉太憑着對烹飪的熱誠,勇於實踐天賦,成為專長的職業,在聖公會聖匠堂教授再培訓課程及陪月課程;於屯門明愛僱員再培訓中心、社會福利署長沙灣中心、香港遊樂場協會及香港明愛社區及高等教育服務等多間社區中心擔任烹飪班導師。

葉太授課時講解生動,以簡易的做法炮製特色美食,透過食材的運用及配搭,讓學生掌握材料的特點,體驗烹飪的樂趣,桃李滿門。

菜餚會摯友

葉太性情和善,樂意與別人分享美食,經常在家款待良朋摯友,為大家炮製各款令人嘖嘖稱讚、色香味俱全的菜餚。友人邊品嘗美餚,邊輕談淺酌,滿室歡聲笑語。

歡迎加入 Forms Kitchen「滋味會」

登記成為「滋味會」會員
• 可收到最新的飲食資訊 •
• 書展 "驚喜電郵" 優惠 * •
• 可優先參與 Forms Kitchen 舉辦之烹飪分享會 •
• 每月均抽出十位幸運會員，可獲精選書籍或禮品 •
* 幸運會員將會收到驚喜電郵，於書展期間享有額外購書優惠

• 您喜歡哪類飲食叢書？(可選多於 1 項)
□中菜　□西菜　□點心　□烘焙　□湯飲　□甜品　□其他＿＿＿＿＿＿

• 您對哪類飲食題材感興趣，而坊間未有出版品提供，請說明：
＿＿＿＿＿＿＿＿＿＿＿＿＿＿＿＿＿＿＿＿＿＿＿＿＿＿＿＿＿＿＿＿＿

• 此書吸引您的原因是：(可選多於 1 項)
□興趣　　　　□內容豐富　　　□封面吸引　　　□工作或生活需要
□作者　　　　□價錢相宜　　　□其他＿＿＿＿＿＿＿＿＿＿＿＿＿

• 您從何途徑擁有此書？
□書展　　　　□報攤 / 便利店　□書店 (請列明：＿＿＿＿＿＿＿＿＿)
□朋友贈予　　□購物贈品　　　□其他＿＿＿＿＿＿＿＿＿＿＿＿＿

• 您覺得此書的價格：
□偏高　　　　□適中　　　　　□因為喜歡，價錢不拘

• 除食譜外，您喜歡閱讀哪類書籍？(可選多於 1 項)
□玄學　　　□旅遊　　　□心靈勵志　□健康美容　□語言學習　　□小說
□兒童圖書　□家庭教育　□商業創富　□文學　　　□宗教
□其他＿＿＿＿＿＿＿＿＿＿＿＿＿＿＿＿＿＿＿＿＿＿＿＿＿＿＿＿＿

• 您是否有興趣參加作者的烹飪分享活動？
□有興趣　　　　□沒有興趣

• 哪位作者的烹飪分享活動您會有興趣參加？
＿＿＿＿＿＿＿＿＿＿＿＿＿＿＿＿＿＿＿＿＿＿＿＿＿＿＿＿＿＿＿＿＿

姓名：＿＿＿＿＿＿＿＿＿＿＿＿□男 / □女　　□單身 / □已婚

聯絡電話：＿＿＿＿＿＿＿＿＿　電郵：＿＿＿＿＿＿＿＿＿＿＿＿＿＿＿

地址：＿＿＿＿＿＿＿＿＿＿＿＿＿＿＿＿＿＿＿＿＿＿＿＿＿＿＿＿＿

年齡：□ 20 歲或以下　　　□ 21-30 歲　　　□ 31-45 歲　　　□ 46 歲或以上

職業：□文職　　　□主婦　　　□退休　　　□學生　　　□其他＿＿＿＿＿

填妥資料後可：
寄回：香港鰂魚涌英皇道 1065 號東達中心 1305 室「Forms Kitchen」收
或傳真至：(852) 2565 5539
或電郵至：marketing@formspub.com

100道鹹甜點心 | 100 Dim Sum & Snacks

作者　Author
葉胡影儀

策劃/編輯　Project Editor
Karen Kan

攝影　Photographer
Imagine Union

美術統籌　Art Direction
Ami

美術設計　Design
Man

出版者　Publisher
Forms Kitchen,
an imprint of Wan Li Book Company Ltd.
香港鰂魚涌英皇道1065號東達中心1305室　Rm 1305, Eastern Centre, 1065 King's Road, Quarry Bay, Hong Kong
電話　Tel: 2564-7511
傳真　Fax: 2565-5539
電郵　Email: marketing@formspub.com
網址　Web Site: http://www.formspub.com
　　　http://www.facebook.com/formspub

發行者　Distributor
香港聯合書刊物流有限公司　SUP Publishing Logistics (HK) Ltd.
香港新界大埔汀麗路36號　3/F., C&C Building, 36 Ting Lai Road,
中華商務印刷大廈3字樓　Tai Po, N.T., Hong Kong
電話　Tel: 2150 2100
傳真　Fax: 2407 3062
電郵　Email: info@suplogistics.com.hk

承印者　Printer
中華商務彩色印刷包裝有限公司　C & C Offset Printing Co., Ltd.

出版日期　Publishing Date
二〇一五年十月第一次印刷　First print in October 2015

瀏覽網站

會員申請